Lecture Notes in Mathematics

Volume 2324

This series reports on new developments in all areas of mathematics and their applications - quickly, informally and at a high level. Mathematical texts analysing new developments in modelling and numerical simulation are welcome. The type of material considered for publication includes:

1. Research monographs
2. Lectures on a new field or presentations of a new angle in a classical field
3. Summer schools and intensive courses on topics of current research.

Texts which are out of print but still in demand may also be considered if they fall within these categories. The timeliness of a manuscript is sometimes more important than its form, which may be preliminary or tentative. Please visit the LNM Editorial Policy (https://drive.google.com/file/d/19XzCzDXr0FyfcV-nwVojWYTIIhCeo2LN/view?usp=sharing)

Titles from this series are indexed by Scopus, Web of Science, Mathematical Reviews, and zbMATH.

Petra Schwer

CAT(0) Cube Complexes

An Introduction

 Springer

Petra Schwer (iD)
Department of Mathematics
Otto-von-Guericke University Magdeburg
Magdeburg, Germany

ISSN 0075-8434   ISSN 1617-9692   (electronic)
Lecture Notes in Mathematics
ISBN 978-3-031-43621-5   ISBN 978-3-031-43622-2   (eBook)
https://doi.org/10.1007/978-3-031-43622-2

Mathematics Subject Classification: Primary: 20F65, 20F67, Secondary: 51F99, 05E18, 20E08, 20F55 , 20F65, 53C23

This Springer imprint is published by the registered company Springer Nature Switzerland AG
The registered company address is: Gewerbestrasse 11, 6330 Cham, Switzerland

Paper in this product is recyclable.

Für Oma.
Wir haben noch immer nicht alles
ausgerechnet.

Preface

This book . . .

. . . provides an accessible introduction to the beautiful theory of CAT(0) cube complexes, without assuming prior knowledge of non-positively curved metric spaces, geometric group theory and algebraic topology. The material begins with a basic introduction to the geometry of non-positively curved metric spaces and the CAT(0) property. A proof of Gromov's characterization of the CAT(0) condition in the class of polyhedral and, more precisely, cube complexes is included. Further topics include the hyperplane and half-space structure of a cube complex, the construction of the Roller boundary and a detailed explanation of the group cubulation method due to Sageev. In addition, further structural properties, some algebraic consequences for groups with geometric actions on CAT(0) cube complexes and some applications are discussed. These include the Tits alternative, special cube complexes and fixed-point properties.

. . . is aimed at advanced undergraduate and early graduate students, who want to learn about a current and vivid research topic. It can also serve as a tool for a reading course or self-study for graduate students and researchers of adjacent areas. It contains about as much content as needed for a one-semester course meeting four hours per week.

. . . does cover a variety of topics but does not aim to serve as an up-to-date research compendium on cube complexes. It does not cover many of the more recent developments and topics in the area. The main focus in writing this book was to provide an accessible introduction for a wide audience at an elementary level. For this reason, many beautiful theorems had to be omitted as their proofs are more involved or rely on material and prerequisites not available to the targeted audience.

. . . grew out of my lecture notes of a one-semester course on CAT(0) cube complexes, which I taught in Münster during the winter term 2012/2013 and in Karlsruhe during the winter term 2014/2015. An earlier version was tested in a reading course with students in Magdeburg during the summer of 2022. Most of the students attending these courses did not have a strong background in metric geometry or in geometric group theory. This forced me to keep the material as elementary and self-contained as possible. However, a basic knowledge of universal covers and fundamental groups is helpful and assumed in (very) few spots throughout the book—I have tried to keep their use to a minimum. If lecturing the course, one might consider explaining the concept of a universal cover when first used.

... contains seven chapters. Each of the chapters comes with plenty illustrations a selection of exercises. Below, I outline the interdependence of chapters and sections and explain which sections could be skipped without risk.

- Section 3.2 can be omitted entirely.
- One can skip parts of the material in Sect. 4.1 and directly introduce cube and all-right spherical complexes as done in Sect. 4.2.
- If the course needs to be drastically shorter, one can skip all of the Chaps. 2 and 3 and define the CAT(0) property for cube complexes only using Gromov's link condition as stated in Theorem 4.43.
- The material of Sect. 4.3 is only used in Sect. 7.3, in particular for the proof of Proposition 7.73. If the latter section is not covered, the material in Sect. 4.3 is also not needed.
- Section 4.4 uses the universal cover of a polyhedral complex. This concept is not explained in the book.
- Chapter 6 heavily relies on Sect. 5.2. Within Chap. 6, there is a some flexibility: from 6.2 one only needs Definitions 6.12 and 6.14 as well as Lemma 6.13. The rest of the material in this section can easily be skipped.
- The sections of Chap. 7 can be read or taught in any order or even skipped entirely. Section 7.1 makes use of material introduced in Sect. 5.1 and of the Bruhat–Tits fixed point theorem (Theorem 3.15) and other bits and pieces of Sect. 3.1. Finally, Sect. 7.3 relies on Sects. 5.1 and 4.3.

... benefited from the help of others. I would like to thank the students of the aforementioned courses for their interest in the topic. Many thanks to Anna Wienhard for insisting that the notes [Sch19] should be turned into a (this) book. I owe thanks to Noam von Rotberg and Frederik Horn who produced almost all of the figures. Noam carefully edited all of them to address the needs of color-blind readers and to facilitate black and white printing. I am grateful to Clara Löh, Joshua Maglione, Yuri Santos Rego and Olga Varghese for many valuable suggestions and comments and to Isobel Davies, Caspar Heusinger, Franziska Hofmann, Joshua Maglione, Yuri Santos Rego, Bakul Sathaye, Matthias Uschold and Noam von Rotberg for catching some typos or proofreading parts of earlier versions of this text.

Finally, I am indebted to the three referees for their thoughtful and detailed reports. A big thank you goes out to my husband for his love and ongoing support and to my kids, who patiently tolerated me spending many odd hours editing this text.

. . . probably still contains typos or errors. If you find some, I would appreciate if you let me know. I plan to post a list of corrections on my website.

Magdeburg, Germany   Petra Schwer
March 2023

Contents

Chapter 1
Introduction

One of the guiding themes in geometric group theory is understanding groups by viewing them as symmetries of well-behaved and easy-to-study spaces. Among such spaces, CAT(0) cube complexes play a significant and successful role. Their metric and combinatorial structure give rise to several nice algebraic properties for groups acting geometrically, that is, properly and cocompactly, on them. The existence of such a cocompact cubulation[1] of a group G implies, for instance, that G is biautomatic [NR03], satisfies rank rigidity [CS11], has finite asymptotic dimension [Wri12] and satisfies the Tits alternative [SW05].

Thus, an effective way to study the algebraic behavior of a group is to cubulate it, that is, to construct a CAT(0) cube complex on which the group acts appropriately. Sageev [Sag95] provides general criteria for when to obtain such an action. The class of groups that admit isometric (or even proper and cocompact) actions on a CAT(0) cube complex is surprisingly large. It includes many prominent classes of finitely generated infinite groups, including (right-angled) Coxeter groups [CD95, NR03], many Artin groups [BM00, PW14, HJP16], a large class of hyperbolic 3-manifold groups [Wis12], some lattices in $SO(n, 1)$ [BHW11], hyperbolic free-by-cyclic groups [HW15, HW16], groups satisfying sufficiently strong small-cancellation conditions [Wis04], random groups at small density in Gromov's model [OW11] and many more.

The study of CAT(0) cube complexes also had a significant impact on low-dimensional topology. In [Ago13], Agol showed that all cubulated hyperbolic groups are virtually special—as conjectured by Wise [Wis21]. Work of Bergeron and Wise [BW12] provides cubulations of 3-manifold groups. Combining this with Agol's theorem one obtains a positive solution to Thurston's long standing virtual Haken and virtual fibering conjectures in 3-manifold theory. More on this in Sect. 7.3.3. Special cube complexes have more refined combinatorial properties

---

[1] By a *cubulation* of a group we mean a proper action on a CAT(0) cube complex. Such a cubulation is *cocompact* if, in addition, the action of the group on the complex is cocompact.

than general cube complexes. Their hyperplanes and intersections of hyperplanes are particularly well behaved. Hence, groups acting geometrically on special cube complexes satisfy stronger algebraic properties than groups "just" acting on CAT(0) cube complexes. One obtains, for instance, that a group that is virtually the fundamental group of a finite special cube complex has separable subgroups and is residually finite, $\mathbb{Z}$-linear and conjugacy separable [HW08, Min12].

1.1   What Is a Cube Complex?

A cube complex is a space built from cubes of arbitrary dimensions glued together along their faces. More precisely, it is a metric cell complex constructed as a quotient space from a disjoint union of Euclidean unit cubes $[0, 1]^n$ of dimension n, where faces of the same dimension $k < n$ belonging to distinct cubes are glued together by isometries.

As illustrated in Fig. 1.1, the standard tiling of Euclidean n-space by unit cubes provides the first examples of cube complexes. More generally, any simplicial graph is a cube complex.

In this book we focus on finite dimensional cube complexes X, which means that the supremum of the dimensions of all cubes in X is finite. We equip X with the metric induced by the Euclidean metric on the individual cubes.

Turning a cube complex X into a metric space, allows us to study curvature properties of X and its impact on groups acting by isometries on the complex. This brings us to the notion of a CAT(0) space, which we now examine.

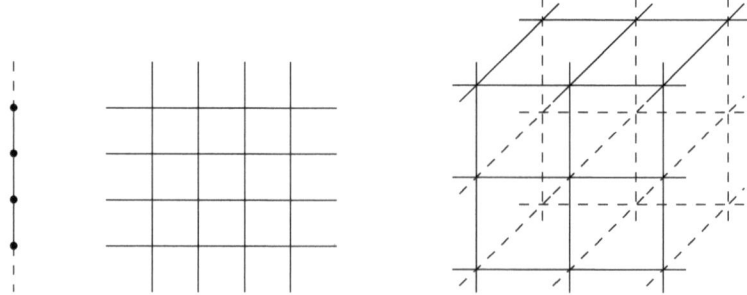

Fig. 1.1 Euclidean n-space naturally carries the structure of a cube complex. This figure shows the standard cubulation for $n = 1, 2$ and 3

1.2   What Is the CAT(0) Property?

A CAT(0) space is a non-positively curved metric space in the sense of comparison geometry. To be a bit more precise, it is a complete geodesic metric space in which geodesic triangles are at most as thick as in Euclidean space.

Notions of non-positive curvature for general path-metric spaces were first studied in the work of Cartan, Busemann, Alexandrov, and Toponogov in the 1950s. Gromov then introduced the term CAT(0) (in honor of Cartan, Alexandrov and Toponogov) for a curvature criterion applicable to a wide class of metric, and metrized polyhedral spaces. He showed that many properties of manifolds with non-positive sectional curvature also hold in this more general setting.

CAT(0) spaces have proven to be useful in geometric group theory. This class of spaces contains many prominent examples, such as symmetric spaces of non-compact type, Riemannian manifolds with sectional curvature uniformly bounded above by 0, trees, (Bruhat)-Tits buildings and many other interesting piecewise Euclidean cell complexes. The CAT(0) cube complexes form a particularly well studied class of such spaces. One of the reasons is that the CAT(0) property is fairly easy to verify in this class thanks to its combinatorial characterizations via links.

The CAT(0) (and more generally CAT(κ)) property is a condition on geodesic triangles in a geodesic metric space. It measures the curvature of a geodesic metric space by comparing its triangles with geodesic triangles of same side lengths in the Euclidean plane (or more generally, the simply connected Riemannian manifold of constant curvature κ). A space is CAT(0) if all its triangles satisfy the CAT(0) condition.

See Fig. 1.2 for an illustration of the CAT(0) comparison condition for triangles in a geodesic metric space. Here, Δ is a triangle in some space and $\bar{\Delta}$ its comparison triangle in the Euclidean plane. Sides with the same marking are assumed to have the same length. The CAT(0) condition is satisfied by Δ if the distance of all points m, n on the triangle Δ is not greater than the distance of the corresponding points $\bar{m}, \bar{n}$ on $\bar{\Delta}$.

Introductions to general CAT(0) theory and consequences for groups acting geometrically on them can be found in the wonderful textbooks [BGS85, Bal95, BH99] and [BBI01].

Fig. 1.2 Given a triangle Δ in some metric space (X, d) and its comparison triangle $\bar{\Delta}$ in Euclidean space. The CAT(0) condition asks that $d = d_X(p, q) \leq \bar{d} = d_\kappa(\bar{p}, \bar{q})$

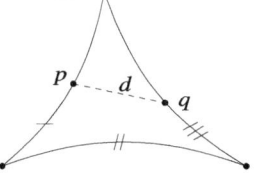
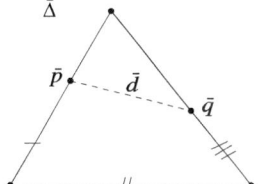

1.3   Which Cube Complexes Are CAT(0)?

In general, the CAT(0) property is hard to test for—remember that one has to compute the diameters of all triangles in a space. Only a very limited number of methods are available. However, within the class of cube complexes this condition becomes purely combinatorial. Being locally CAT(0) can be described using Gromov's link condition within the class of polyhedral spaces. But, for cube complexes equipped with the intrinsic length metric, the link condition turns into an easy-to-check combinatorial criterion: a simply connected cube complex, all of whose vertex links are flag simplicial complexes, is CAT(0). See [Gro87] or Theorem II.5.20 in [BH99] for a proof in the finite dimensional case and [Lea13] for a proof in the general case.

Examples of CAT(0) cube complexes include trees, Euclidean spaces and right-angled buildings. All of these classes have a certain tree-like behavior. Certain Cayley graphs and presentation complexes provide examples of CAT(0) cube complexes. See Fig. 1.3 for the following two examples. The Cayley graph $\Gamma(F_2, \{a, b\})$ of the free group on two generators is a regular 4-valent tree and hence CAT(0). The presentation complex spanned on the Cayley graph $\Gamma(\mathbb{Z}^2, \{a, b\})$ of $\mathbb{Z}^2$ with standard generators is the standard tiling of the Euclidean plane by unit squares.

CAT(0) cube complexes also arise naturally in the context of right-angled Artin and Coxeter groups as we see in this book. More on this topic can be found, for instance, in [Wis12, Sag14] or [Wil17].

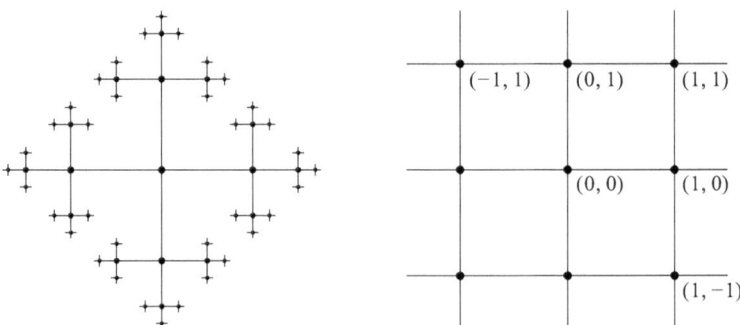

Fig. 1.3 The Cayley graphs $\Gamma(F_2, \{a, b\})$ on the left and $\Gamma(\mathbb{Z}^2, \{a, b\})$ on the right are cube complexes

1.4   Why Should One Study CAT(0) Cube Complexes?

Upper curvature bounds on spaces have a big impact on the algebraic properties of a group acting geometrically on such spaces. The easy to check characterization of the CAT(0) property makes cube complexes the ideal setting to study non-positive curvature with combinatorial methods.

The geometry of CAT(0) cube complexes is much better understood than that of general CAT(0) spaces. The combinatorial nature gives them a tree-like structure and many groups admit a non-trivial action on them with strong algebraic consequences. For instance, groups acting properly on finite dimensional CAT(0) cube complexes are known to have finite asymptotic dimension [Wri12]. The structure added to the space by its hyperplanes (more on them below) allows one to prove that the Tits alternative holds [CS11, SW05]. Whether all CAT(0) groups satisfy the Tits alternative remains an open problem at this point.

Moreover, there are connections to other subjects. I would like to emphasize Kazhdan's Property (T), see e.g. [CDH10, NR98], and the connection to median spaces, see [Che00, Rol98]. The 1-skeleton of a CAT(0) cube complex is a median graph when endowed with its intrinsic path-metric. This means that for any three points, there exists a unique point that lies between any two of them. This fact is closely related to the existence of hyperplanes and allows for a graph theoretic approach that can not be generalized to all CAT(0) spaces.

A powerful tool in studying CAT(0) cube complexes is the combinatorics of their hyperplanes. A hyperplane is a closed convex subspace of a CAT(0) cube complex that consists of midcubes of cubes, which itself carries a cubical structure of one dimension lower. Its complement has exactly two components, both of which are convex. The closures of these components are called half-spaces of the cube complex. In a CAT(0) cube complex the midpoint of each edge in the 1-skeleton belongs to a unique hyperplane.

Understanding hyperplanes and their relative positions, such as how they intersect and interact, helps one to study group actions on CAT(0) cube complexes. See Fig. 1.4 for some first examples of hyperplanes in a cube complex.

Hyperplanes are particularly well behaved for special cube complexes—a class of spaces introduced by Haglund and Wise [HW08]. Their study is part of Wise's *cubical route to understanding groups* as outlined in [Wis14]. The main strategy is the following: to understand a group G first get G to act properly and cocompactly on a CAT(0) cube complex G. Then show that the quotient of the space by the action is special up to taking finite covers. In the last step apply work of Haglund and Wise [HW08] to deduce many algebraic consequences for G. Cube complexes in general and this strategy in particular had a huge impact on the development of geometric group theory in recent years.

In the case when G is the fundamental group of a closed 3-manifold M this strategy was used to obtain solutions to Thurston's virtual fibering and virtual Haken conjectures. See [Ota14] for a statement of these conjectures. By work of Kahn and Markovic [KM12] and Bergeron and Wise [BW12] there are enough walls (obtained

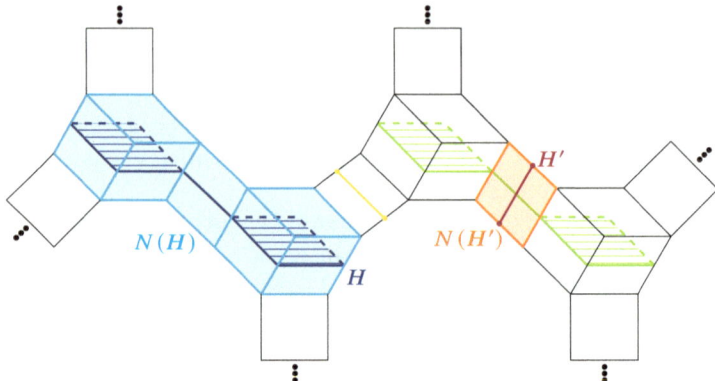

Fig. 1.4 This figure shows a piece of the CAT(0) cubulation of $PGL_2(\mathbb{Z})$ with some of its hyperplanes. There are two types of hyperplanes. The hyperplanes in one class consist of a single edge, like H', while the others consist of two squares glued to the vertices of an edge, see H. See Example 5.5 for further details

from M) to construct a CAT(0) cube complex on which G acts geometrically. This cube complex X is constructed using Sageev's cubulation method [Sag95], as explained in Sect. 5.2. Finally Agol [Ago13] proved that the quotient X/G is special. His result was the final missing piece to deduce the virtual fibering and virtual Haken conjectures—closing long-standing open problems in 3-dimensional topology. I give a more detailed account on this in Sect. 7.3.3. See also Agol's ICM notes on this topic [Ago12].

Boundaries of non-positively curved metric spaces provide tools to study their large scale behavior, local-to-global phenomena and the dynamics at infinity.

The Roller boundary [Rol98] is a compactification of a CAT(0) cube complex constructed by means of filters of half-spaces in the cube complex. Nevo and Sageev [NS13] show that if X is a locally finite, well-behaved cube complex and admits a cocompact action by a group G, then a subset $B(X)$ of the Roller boundary is the realization of the Poisson boundary. Levcovitz [Lev21] provides an equivalent characterization for $B(X)$ being equal to the entire Roller boundary. He uses this to prove that the Roller boundary of the usual CAT(0) cube complex of a right-angled Artin group is equal to $B(X)$, as long as the corresponding group does not decompose as a direct product. Beyrer, Fioravanti and Incerti-Medici [BFIM21] prove that one can reconstruct a cube complex from a cross-ratio on its Roller boundary. This generalizes results known for symmetric spaces and Bruhat-Tits buildings.

As already mentioned at the beginning of this introduction, a large class of groups admit interesting and even geometric actions on CAT(0) cube complexes with a strong impact on their algebraic behavior. There are many more properties than just the ones mentioned so far.

One could, for instance, ask about fixed points of actions. It turns out that a finitely generated group acts on a CAT(0) cube complex without a global fixed point if and only if it has a codimension one subgroup [Sag95, Ger98]. Caprace and Sageev prove that any group acting essentially fixed-point free at infinity on an irreducible CAT(0) cube complex of finite dimension contains a rank one isometry. As a consequence, they obtain rank rigidity and a geometric proof of the Tits alternative for CAT(0) cube complexes.

1.5 Where to Look for Cube Complexes?

Some ad-hoc examples of CAT(0) cube complexes were provided above and include trees, products of trees, Euclidean spaces and right-angled buildings. More examples arise naturally in the context of right-angled Artin or Coxeter groups as presentation- or Salvetti complexes. Niblo and Reeves [NR03] show how to construct a CAT(0) cube complex for every finitely generated Coxeter group. If the group is word-hyperbolic, the action on this complex is cocompact.

An influential application of CAT(0) cube complexes, due to Billera, Holmes and Vogtmann [BHV01], is the fact that the space of phylogenetic trees is a CAT(0) space. The ways in which species evolve or languages change can be modeled using trees. These trees, in turn, form a space carrying a natural CAT(0) cubical structure.

There are various other structures in metric geometry and graph theory that are closely related to CAT(0) cube complexes. For example, median graphs and algebras, Helly graphs and their connection to median and clique graphs, (weakly) modular and bridged graphs and solution sets to the 2-SAT problem in computer science, just to mention a few. A detailed account of such structures can be found in the survey by Bandelt and Chepoi [BC08].

Lastly, the nice algorithmic properties of CAT(0) cube complexes allow for many non-group theoretic applications. It is, for instance, relatively easy to compute geodesics in a CAT(0) cube complex as Ardila, Owen and Sullivant show [AOS12]. Moreover, every CAT(0) cube complex may be realized as a state complex of a reconfigurable system. Understanding how CAT(0) cube complexes can be used in robot motion planning is an active research problem covering a broad range of disciplines [AOS12, ABY14, AG04]. Given, for instance, a robot moving on a grid, the configuration space of possible positions of the robot forms a cube complex. When this space is CAT(0), one can explicitly construct shortest paths between possible positions of the robot. See also [AM20] for a survey on this topic.

1.6 What's in This Book?

This book provides an accessible introduction to the basic notions of CAT(0) metric spaces and contains the proof of Gromov's combinatorial characterization of the CAT(0) property in the class of cube complexes. We provide insight into the hyperplane and half-space structure of a cube complex, explain how to construct the

Roller boundary and give a detailed explanation of the group cubulation method due to Sageev. In addition, further structural properties, some algebraic consequences for groups with (geometric) actions on CAT(0) cube complexes, as well as some applications are discussed.

Chapter 2 starts with a brief introduction to the theory of metric spaces in its first section. A guiding principal of geometric group theory is to view a group as a metric space. How this can be achieved is discussed in Sect. 2.2.

What is a CAT(0) metric space? This question is answered in Chap. 3. In Sect. 3.1 some general properties and examples of metric spaces of non-positive curvature are discussed and the CAT(0) property is introduced. Sections 3.2 and 3.3 then contain a characterization of CAT(0) spaces using angles and the notion of a flat cone over a polyhedral complex.

Cube complexes and Gromov's combinatorial characterization of non-positive curvature in this setting are then discussed in Chap. 4 answering the question of which cube complexes are CAT(0). Section 4.1 first introduces the wider class of polyhedral complexes and the link condition constituting non-positive curvature in this class. This broader viewpoint is narrowed down to cube complexes in Sect. 4.2, where the equivalent formulation of the link condition in terms of flag simplicial complexes is also considered. Section 4.3 contains a result due to Haglund and Wise, which explains how a cube complex can be completed to a CAT(0) cube complex. Finally, in Sect. 4.4 a first construction of an interesting class of CAT(0) cube complexes associated with right-angled Artin groups is provided.

Hyperplanes and the associated half-space systems are introduced in Chap. 5. The first two sections, Sects. 5.1 and 5.2, contain the definitions and main properties of hyperplanes and half-spaces. Each cube complex gives rise to a system of half-spaces satisfying certain defining properties. This system is obtained from the collection of all half-spaces associated with walls in the cube complex. It is explained in Sect. 5.2 how one can construct a cube complex from any given (abstract) half-space system. Section 5.3 on Roller duality then discusses the connection between a cube complex and the complex constructed from its half-space system.

As mentioned above, cubulating groups and having them act on a cube complex is a powerful way to study their algebraic structure. How to cubulate a Coxeter group is explained in Chap. 6. Coxeter groups are among the most well studied classes of finitely presented infinite groups due to their appearance in many combinatorial settings in algebra and combinatorial group theory and their role as Weyl groups. We introduce them in Sect. 6.1 by means of generators and relations. Reflections and the deletion condition are discussed in Sect. 6.2. Hyperplanes, walls and half-space systems for Coxeter groups are constructed in Sects. 6.3 and 6.4, which allows one to obtain a CAT(0) cube complex for the Coxeter groups to act on.

Chapter 7, contains a collection of topics, which illustrate the broad and varied theory of CAT(0) cube complexes. Section 7.1 highlights the analogy between cube complexes and trees by studying fixed point properties of group actions. The material in Sect. 7.2 has a more algebraic flavor. This section contains an almost complete proof of the Tits alternative, relying only on a few algebraic black boxes.

Special cube complexes, which play a major role in Agol's proof of the virtual Haken conjecture, are introduced in Sect. 7.3. We also comment on the history and context of Agol's result in the final subsection of that section. We conclude this rainbow of topics with an application of cube complexes in the context of evolutionary biology and show in Sect. 7.4 that the space of phylogenetic trees is a CAT(0) cube complex.

1.7 Some (of the Many) Things I Have Omitted

There are a number of interesting and more recent developments and topics in the area that are not included in this book. For one, the selection of the material is dictated by personal taste. Second, the main priority was to provide an introduction accessible to a wide audience—in particular to researchers and students without a strong background in either geometric group theory or algebraic topology. As a result many theorems had to be skipped since the tools needed to prove them were not available.

Among the omitted material is the cubical small-cancellation theory [Wis04, Wis12, Wis21] and a proof of Wise's malnormal special quotient theorem. This theorem is one of the central tools in the theory developed by Wise and Agol. I have also not included a full proof of Agol's theorem, which lead to a solution of the virtual Haken conjecture [Ago13, Ago12]. However, the reader should be well-equipped to read the account on this result provided by Sheperd [She21a] or the original works by Agol and Wise.

Also not covered is Huang's use of cube complexes in studying the quasi-isometric rigidity of right-angled Artin groups. Huang and Kleiner [HK18] characterize groups quasi-isometric to a right-angled Artin group with finite outer automorphism group. All such groups admit a geometric action on a CAT(0) cube complex. Their characterizations were then used by Huang to give a commensurability classification of the groups quasi-isometric to certain right-angled Artin groups [Hua18].

I also wish to draw the reader's attention to the recent solution of the marked length spectrum rigidity problem by Beyrer and Fioravanti [BF22]. Along the way they also describe the relationship between various kinds of boundaries of a CAT(0) cube complex, e.g. the Roller and Gromov boundary.

Stallings introduced folding of morphisms of graphs and used them to represent finitely-generated subgroups of a finite-rank free group as immersions of finite graphs. This construction yields answers to many algorithmic questions in this context. Recently Dani and Levcovitz [DL21] used Stallings-like methods to study subgroups of right-angled Coxeter groups, which act geometrically on CAT(0) cube complexes. Their results where generalized by Ben-Zwi, Kropholler and Lyman [BZKL22].

Generalizations of many results on cube complexes to median or coarse median spaces were obtained. For instance the generalization of the Tits alternative to median spaces by Fioravanti [Fio18] or the generalizations of the coarse geometry of cube complexes by Behrstock, Hagen and Sisto [BHS17]. Fioravanti, Levcovitz and Sageev [FLS22] show that for many right-angled Artin and Coxeter groups, all cocompact cubulations induce the same coarse median structure on the group. With this they provide the first examples of non-hyperbolic groups with this property.

This is by no means an exhaustive list. However, . . .

1.8   Final Remarks

I hope it has become clear that the material in this book begins elementary enough to be accessible to many. At the same time, it opens up the door to a wide variety of topics and a vast world of mathematics in which there is still plenty to explore and discover.

Chapter 2
Metric Spaces Meet Groups

Studying the close connection between groups and geometry is a central concept in geometric group theory. Groups are equipped with metrics themselves or are viewed as symmetries of geometric and combinatorial structures. Some of these structures, such as the Cayley graphs or presentation complexes of a group, are directly constructed from the group itself. This chapter contains a concise introduction to metric spaces and collects the prerequisites for the main body of the book. Cayley graphs are constructed and their main properties are listed and illustrated with examples.

The book by Burago, Burago and Ivanov [BBI01] or Bridson and Haefliger's monograph [BH99] are good sources to read more about the geometry of metric spaces. For a general introduction to geometric group theory we refer to Löh [Lö17].

2.1 A Tiny Bit of Metric Geometry

Metric geometry approaches geometry based on the notion of a length space. One aim is to develop properties of differentiable manifolds solely based on the underlying metric structure of the space. In the last years this theory influenced many other areas of mathematics, including group theory.

This section collects some basic facts about length spaces and geodesic metric spaces and provides a foundation for the remainder of the book.

Definition 2.1 (Metrics) A *metric* on a non-empty set X is a function $d : X \times X \to \mathbb{R}$ such that for all $x, y, z \in X$ one has

1. (positivity) $d(x, y) \geq 0$ with equality if and only if $x = y$,
2. (symmetry) $d(x, y) = d(y, x)$, and
3. (triangle inequality) $d(x, z) \leq d(x, y) + d(y, z)$.

P. Schwer, *CAT(0) Cube Complexes*, Lecture Notes in Mathematics 2324,
https://doi.org/10.1007/978-3-031-43622-2_2

A function $d : X \times X \rightarrow \mathbb{R}$ is a *pseudo-metric* on X if it satisfies symmetry, the triangle inequality and in addition $d(x, x) = 0$.

The difference between a pseudo-metric and a metric is that positivity may not be satisfied for pseudo-metrics. It is then possible that the distance between distinct points is zero.

Definition 2.2 (Metric Spaces) A *metric space* (X, d) is a set X together with a metric d. A metric space X is *complete* if every Cauchy sequence in X converges. For subsets $Y \subset X$ the restriction of the metric d to $Y \times Y$ is referred to as the *restricted* metric on Y.

Denote the open ball $\{y \in X \mid d(x, y) < r\}$ of radius r around $x \in X$ by $B_r(x)$ and its closure by $\overline{B}_r(x) = \{y \in X \mid d(x, y) \leq r\}$.

Basic examples for metric spaces that are not manifolds can be obtained from graphs. In the following we will only consider undirected, simplicial graphs without loops, which are defined as follows:

Definition 2.3 (Graphs) A *graph* is a pair $\Gamma = (V, E)$ of sets, where E is a set of subsets of V containing exactly two elements. That is

$$E \subset \{e \mid e \subset V, |e| = 2\}.$$

The elements of E are called *edges* and the elements of V *vertices* of the graph Γ.

We say that two vertices u, v are *neighbors* or *adjacent* if and only if they are connected by an edge, that is $\{u, v\} \in E$. The *degree* of a vertex is the number of its neighbors, or, equivalently, the number of edges containing e.

We review some basic terminology concerning graphs.

Definition 2.4 (Edge-Paths in Graphs) Let $\Gamma = (V, E)$ be a graph. An *edge-path* in Γ is a sequence of vertices $u_0, u_1, \ldots, u_k$ such that any two subsequent ones are connected by an edge, i.e. $e_i = \{u_{i-1}, u_i\} \in E$ for $i = 1, \ldots, k$. We say that an edge-path *connects* the vertices u and v if $u = u_0$, $v = u_k$ and $k < \infty$. We then refer to k as the *length* of the path. The graph Γ is *connected* if every pair of vertices in Γ is connected by an edge-path.

One can use edge-paths to define a metric on the vertices of a graph.

Example 2.5 (Discrete Metric Spaces via Graphs) Let $\Gamma = (V, E)$ be a connected graph with vertex set V and edge set E. Define a distance function $d : V \times V \rightarrow \mathbb{R}$ on the set of vertices of the graph as follows: for every pair of distinct vertices $u, v \in V$ the distance $d(v, u)$ is defined to be the smallest k for which there exists an edge-path of length k connecting u and v. If $u = v$ we define their distance $d(u, v) = 0$. One can prove that this distance function d satisfies all the properties of a metric on V.

A source of interesting non-discrete examples of metric spaces are metric realizations of graphs. Roughly speaking a *metric realization of a graph* is a graph where each edge is realized by an interval of a certain length.

Definition 2.6 (Metric Realization of Graphs) Let $\Gamma = (V, E)$ be a connected graph. Let $l : E \to [0, \infty)$ be a map that assigns to every edge a non-negative number $l(e)$, its *length*. From Γ and l we obtain a *metric realization* of Γ, which we denote by $|\Gamma|$, as follows: For every edge $e = \{u, v\} \in E$ fix an orientation and write (u, v) for the oriented edge. Take an interval $I_e = [0, l(e)]$ of length $l(e)$ and label the endpoints of the interval with the vertices of e. So if $e = (u, v)$, then $0 \in I_e$ is labeled by u and $l(e) \in I_e$ is labeled by v. Let $\mathcal{E}$ denote the collection of these intervals. Define an equivalence relation on the endpoints of the intervals in $\mathcal{E}$ by letting ends of I_e and $I_{e'}$ be equivalent if both are labeled by the same vertex $u \in V$. The metric realization $|\Gamma|$ of Γ with respect to l is then the quotient of $\mathcal{E}$ by this equivalence relation.

The metric d_Γ on $|\Gamma|$ is defined as follows: Let p, q be two points in the metric realization of Γ. Then $d_\Gamma(p, q) = d_{I_e}(p, q)$ if p, q are both contained in the image of I_e in the quotient. In case p, q are vertices put $d_\Gamma(p, q) = d(p, q)$ as defined in Example 2.5. In case p, q are points not contained in a common edge let I_e be an edge containing p and I_f be an edge containing q. Then put

$$d_\Gamma(p, q) = \min\{d_{I_e}(p, u) + d(u, v) + d_{I_f}(q, v) \mid u \in e \text{ and } v \in f\}.$$

Again $d(u, v)$ is as in Example 2.5.

The details of this folklore construction have been worked out in [Mug21]. Note that Definition 2.6 is an ad-hoc definition of a quotient of metric spaces together with the quotient metric in this particular setting of graphs. See [BBI01] for the general theory on quotient spaces and their metrics.

Remark 2.7 If we do not specify a length function on E, all edges are assumed to have length one, and we consider the geometric realization with respect to the constant function $l(e) = 1$ for all $e \in E$.

By abuse of notation we sometimes simply write Γ instead of $|\Gamma|$.

One could also define a metric on $|\Gamma|$ by considering continuous curves $\gamma : I \to |\Gamma|$, subdividing the defining interval into smaller sub-intervals such that the image of the restriction of γ to the sub-intervals is inside a single edge. Ideas similar to that in Definitions 2.9 and 2.13 will then allow to define a metric on $|\Gamma|$. The two definitions will be equivalent.

Items 2 and 3 of Example 2.8 make use of the concept of a Cayley graph as defined in Definition 2.24.

Example 2.8

1. Let Γ be a connected graph. Consider its metric realization where all edges are assumed to have length one. Then $|\Gamma|$ is a geodesic metric spaces. It is uniquely

geodesic if and only if the underlying combinatorial graph does not have any circuits.
2. If $\Gamma = \Gamma(\mathbb{Z}, \{1\})$ any metric realization is isomorphic to the real line. The metric realization with all edges of length one produces a metric space isomorphic to the real line where vertices may canonically be identified with the elements of $\mathbb{Z} \subset \mathbb{R}$.
3. Let $\Gamma = \Gamma(\mathbb{Z}^2, \{(1, 0)^T, (0, 1)^T\})$. Then the geometric realization with all edges of length one is isomorphic to square lattice inside the plane $\mathbb{R}^2$ with metric induced by the ℓ_1-metric.

For further examples of metric spaces see Example 2.12. We now turn our attention to curves (of shortest length) in metric spaces and consider metrics that realize the length of a shortest connecting curve.

Definition 2.9 (Length of a Curve) Let (X, d) be a metric space, $I \subset \mathbb{R}$ an interval and $\gamma : I \to X$ a continuous map which we refer to as *path* or *curve*. The length of the curve γ is defined by

$$\ell_d(\gamma) = \sup \sum_{i=0}^{n-1} d(\gamma(t_i), \gamma(t_{i+1})) \in [0, \infty),$$

where the supremum is taken over all $n \in \mathbb{N}$ and all sequences $t_0 \leq t_1 \leq \ldots \leq t_n$ of points in I. A curve γ is *rectifiable* if its length is finite.

An important tool in the study of metric spaces is the notion of geodesics, that is, shortest paths connecting pairs of points in a given space. They are defined as follows. It is an easy computation to check that geodesics in the sense defined below are indeed curves of shortest possible length between a given pair of points.

Definition 2.10 (Geodesics) Let (X, d) be a metric space and x, y points in X. A *geodesic* γ in X from x to y, abbreviated by $\gamma : x \rightsquigarrow y$, is a (continuous) map $\gamma : [0, l] \to X$, with $l \in \mathbb{R}^+$, such that $\gamma(0) = x, \gamma(l) = y$ and such that for all $t, t' \in [0, l]$ one has

$$d(\gamma(t), \gamma(t')) = |t - t'|. \tag{2.1}$$

The *geodesic segment* $\mathrm{seg}_y(\gamma, x)$ *between x and y with respect to γ* is the image of $\gamma : x \rightsquigarrow y$ in X. A *geodesic ray* is a continuous map $\gamma : [0, \infty) \to X$ such that $d(\gamma(t), \gamma(t')) = |t - t'|$ for all $t, t' \in \mathbb{R}^+$, whereas a *geodesic line* is a continuous map $\gamma : \mathbb{R} \to X$ that satisfies $d(\gamma(t), \gamma(t')) = |t - t'|$ for all $t, t' \in \mathbb{R}$. A continuous map $\gamma : [a, b] \to X$ is a *local geodesic* if for all $t \in [a, b]$ there exists $\varepsilon > 0$ such that $\gamma|_{[t-\varepsilon, t+\varepsilon]}$ is a geodesic.

Continuity of γ is an automatic consequence of the defining condition that $d(\gamma(t), \gamma(t')) = |t - t'|$. So in fact one does not need to assume that γ is continuous in that part of the definition above.

Geodesics are length minimizing. Among all curves connecting a pair of points a geodesic has the shortest length. For arbitrary metric spaces geodesics connecting distinct pairs of points may not be unique. There exist metric spaces in which there are many (in fact even infinitely many) geodesics connecting certain pairs of points. Examples are provided after the next definition.

Definition 2.11 (Geodesic Metric Space) A metric space (X, d) is *geodesic* if for all pairs of points $x, y \in X$ there exists a geodesic $\gamma : x \rightsquigarrow y$. In case such a geodesic is unique we say that X is *uniquely geodesic*. The space (X, d) is *r-uniquely* geodesic if and only if for all pairs $x, y \in X$ with $d(x, y) < r$ there exists a unique geodesic $\gamma : x \rightsquigarrow y$.

Example 2.12 Here are some examples of geodesic metric spaces for which we discuss the property of being uniquely geodesic.

1. Denote by $(\mathbb{S}^2, d)$ the unit sphere in $\mathbb{R}^3$ together with its standard round metric d. This is an example of a geodesic, but not uniquely geodesic metric space. Observe that any pair of points $x, y \in \mathbb{S}^2$ can be connected by a geodesic. Such a geodesic is unique only if the distance $d(x, y) < \pi$. Any pair of opposite points at distance π is connected by infinitely many geodesics going halfway around the sphere. So $(\mathbb{S}^2, d)$ is not a geodesic space. It is, however, π-uniquely geodesic in the sense defined above. Similarly any higher-dimensional sphere $(\mathbb{S}^n, d)$ equipped with the round metric is geodesic and π-uniquely geodesic.
2. The Euclidean plane, i.e., $\mathbb{R}^2$ together with the standard Euclidean metric d, is uniquely geodesic. In contrast, let d_1 be the metric on $\mathbb{R}^2$ that is defined using the l_1-norm:

$$d_1(x, y) = |x_1 - y_1| + |x_2 - y_2|.$$

The metric space $(\mathbb{R}^2, d_1)$ is not uniquely geodesic.

Metric spaces in which the distance between pairs of points can always be approximated by rectifiable curves are considered in the next definition.

Definition 2.13 (Length Metric and Length Space) Let (X, d) be a metric space. Its metric d is a *length metric* if

$$d(x, y) = \inf\{\ell(\gamma) \mid \gamma \text{ a rectifiable curve connecting } x \text{ and } y\}.$$

A metric space equipped with a length metric is referred to as a *length space*.

Not every length space is a geodesic space. For example, the set $\mathbb{R}^2 \setminus \{(0, 0)\}$ with the restricted Euclidean metric or the induced length metric is a length space, which is not geodesic. There is, for example, no geodesic connecting a pair of points where one lies on the negative side of the real axis and the other on the positive side.

One can turn any metric space into a length space as the following proposition shows.

Proposition 2.14 (Length Metric) *Let (X, d) be a metric space and define a map* $\bar{d} : X \times X \to [0, \infty]$ *by putting*

$$\bar{d}(x, y) = \{\ell(\gamma) \mid \gamma \text{ a rectifiable curve connecting } x \text{ and } y\}.$$

Then the following are true.

1. *$\bar{d}$ is a metric on X.*
2. *$\bar{d}(x, y) \geq d(x, y)$ for all $x, y \in X$.*
3. *If a curve $c : [a, b] \to X$ is continuous and rectifiable in (X, d), then it is also continuous and rectifiable in $(X, \bar{d})$.*
4. *For rectifiable curves c one has $\ell_d(c) = \ell_{\bar{d}}(c)$.*
5. *$\bar{\bar{d}} = \bar{d}$.*

Proof Items 1 and 2 are easy consequences of the definition of the length of a curve. From Proposition I.1.20 (5) in [BH99] we obtain item 3. We next prove item 4. Let $c : [a, b] \to X$ be a curve of length $\ell_{\bar{d}}(c)$ with respect to $\bar{d}$ and of length $\ell_d(c)$ with respect to d. Item 2 then implies that $\ell_{\bar{d}}(c) \geq \ell_d(c)$. Further

$$\ell_{\bar{d}}(c) = \sup \sum_{i=0}^{k-1} \bar{d}(c(t_i), c(t_{i+1})) \leq \sup \sum_{i=0}^{k-1} \ell_d(c|_{[t_i, t_{i+1}]}) = \ell_d(c),$$

where the supremum is taken over all sequences (t_i) with $t_0 = a, t_k = b$. Finally, item 3 combined with item 4 implies item 5.

Notation 2.15 We call $\bar{d}$ the *length metric associated with d*. For all subsets $Y \subset X$ we refer to the length metric associated with the restriction of d to Y as the *induced length metric* on Y. In general, the induced length metric $\bar{d}|_Y$ on Y differs from the length metric associated with the restriction of d to Y. Compare also Def. 3.3 on page 33 of [BH99] and the examples after it.

Length metrics are well behaved. In particular they induce natural length metrics on metric quotients and metric coverings. See also Theorem 3.18.

Example 2.16 Let X denote the Euclidean plane with the open first quadrant removed, i.e., the space $X := \mathbb{R}^2 \setminus Q_1$, where $Q_1 = \{(x, y) \mid x > 0, y > 0\}$. Let d_l denote the induced length metric on this subset of the Euclidean plane. This metric differs from the restricted Euclidean metric d_e on the same set. In particular (X, d_l) is a geodesic space whereas (X, d_e) is not. Figure 2.1 shows the space X and one of the geodesics with respect to d_l connecting the points labeled x and y in the set. The same pair of points is not connected with a geodesic with respect to the restricted metric. Such a geodesic would have to have the same length as a geodesic in the Euclidean plane (shown as dotted path), which necessarily cuts through Q_1 and is hence not contained in X.

The next theorem concerns the intrinsic geometry of certain length spaces. For a proof see Proposition I.3.7 in [BH99] or Chapter I.2 in [Bal95]. In particular, the

Fig. 2.1 Equip $\mathbb{R}^2 \setminus Q_1$ with the induced length metric by the Euclidean metric on $\mathbb{R}^2$. The (red) geodesic connecting x and y in $\mathbb{R}^2 \setminus Q_1$ is longer than the (dotted red) geodesic in $\mathbb{R}^2$. See Example 2.16 for details

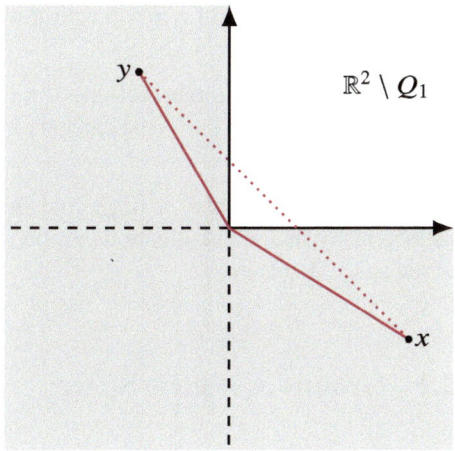

Hopf–Rinow theorem implies, that given any two points $x, y \in X$ there exists a geodesic connecting them.

Theorem 2.17 (Hopf–Rinow Theorem) *Let X be a complete and locally compact length space. Then*

1. *every closed and bounded subset of X is compact. And*
2. *X is geodesic.*

Central to us is the relationship between metric spaces and their symmetry groups. Most often we consider spaces up to isometry. That is, up to bijective transformations of the space that keep distances and hence the overall metric shape of the space intact.

Definition 2.18 (Isometry) An *isometry* $\phi : X \rightarrow X'$ between metric spaces (X, d) and (X', d') is a bijective map, such that

$$d(x, y) = d'(\phi(x), \phi(y)) \text{ for all } x, y \in X.$$

We write $\mathrm{Iso}(X)$ for the group of all isometries $\phi : X \rightarrow X$.

The set $\mathrm{Iso}(X)$ of all isometries of a space X forms a group. This group acts on X in the following sense.

Definition 2.19 (Group Action) Let G be a group, $\mathcal{C}$ a category and $X \in \mathrm{Obj}(\mathcal{C})$. A *group action* of G on X is a homomorphism $G \rightarrow \mathrm{Aut}_\mathcal{C}(X)$, that is, a family of elements $\{f_g \in \mathrm{Aut}(X)\}_{g \in G}$ such that for all $g, h \in G$ one has

$$f_g \circ f_h = f_{gh}.$$

We use the shorthand notation $g.x$ for the image $f_g(x)$ of x.

To see an example of a group action let $\mathcal{C}$ be the category of topological spaces and $\mathrm{Aut}_{\mathcal{C}}(X) = \mathrm{Homeo}(X)$.

Definition 2.20 (Isometric Action) An *isometric action* of a group G on a metric space X is a group action in the category of metric spaces, that is, a homomorphism $\phi : G \to \mathrm{Iso}(X)$.

Isometric actions play a huge role in geometric group theory. They allow to deduce many interesting algebraic properties for a group from geometric properties of the space it acts on.

2.2 Groups as Metric Spaces

A leading theme in geometric group theory is to consider groups themselves as metric spaces. For a finitely generated group there is a simple way to turn the group into a graph, namely the construction of a Cayley graph with respect to a given set of generators. These graphs are naturally endowed with a metric, which encodes a lot of the large scale structure of the group.

In this section we prove the fundamental theorem of geometric group theory as Theorem 2.36. The viewpoint taken in this section sets the theme for the entire book.

We start by constructing groups via generators and relations. For further details see for example Section 2.2.3 in [Lö17].

Definition 2.21 (Normal Generation) Let G be a group and let R a subset of G. The *normal subgroup of G generated by R* is the smallest normal subgroup of G containing R. We denote this group by $\langle R \rangle_G^{\triangleleft}$.

For every subset R in G the normal subgroup of G generated by R can be described as follows:

$$\langle R \rangle_G^{\triangleleft} = \bigcap \{ H \mid H \trianglelefteq G \text{ and } R \subset H \}.$$

Definition 2.22 (Generators and Relations) Let S be a set and let R be a subset of the set of words in $S \cup S^{-1}$. Denote by $F(S)$ the free group generated by S. We say the group

$$\langle S \mid R \rangle := F(S)/\langle R \rangle_{F(S)}^{\triangleleft} \tag{2.2}$$

is *generated by S with respect to the relations R*. The elements in S are the *generators* of G and the words in R are referred to as *relations*.

If a group G is isomorphic to $\langle S \mid R \rangle$ for some S and R we say $\langle S \mid R \rangle$ is a *presentation of G or G is presented by the pair (S, R)*.

As a set, the free group $F(S)$ is the collection of all equivalence classes of finite length words in the generators S and their formal inverses. Two such words are considered equivalent if they can be transformed into one another by deletion or insertion of expressions of the form ss^{-1} for $s \in S \cup S^{-1}$. The identity element in the free group is represented by the empty word. Multiplication of two words is concatenation. Note that the map sending every $s \in S$ to $s \in \langle S \mid R \rangle$ extends to a homomorphism $\phi : F(S) \to \langle S \mid R \rangle$ having $\langle R \rangle_{F(S)}^{\triangleleft}$ as its kernel.

Definition 2.23 (Finitely Generated and Presented Groups) A group G is *finitely generated* if there exists a finite subset $S \subset G$ generating the group. We say G is *finitely presented* if it is finitely generated by a set $S \subset G$ and there exists a finite set R of relations such that G is isomorphic to $\langle S \mid R \rangle$.

One can associate a (labeled) graph to a pair of a group and a generating set.

Definition 2.24 (Cayley Graph) Let G be a group and $S \subset G$ a generating set of G. Suppose that S does not contain the identity element. The *Cayley graph* $\Gamma = \Gamma(G, S)$ of G with respect to S is defined as follows: The set of vertices of Γ is G. A pair $\{g, h\}$ is an edge if there exists $s \in S \cup S^{-1}$ such that $gs = h$.

Note that, by construction, $\Gamma(G, S) = \Gamma(G, S^{-1}) = \Gamma(G, S \cup S^{-1})$.

Remark 2.25 In case the identity element would be contained in a generating set S of a group G the Cayley graph $\Gamma(G, S)$ would contain a loop at every vertex. These loops do not encode any useful (large scale) information about G. We hence assume that the identity is not contained in S.

Sometimes oriented Cayley graphs are used. Then edges are taken to be ordered pairs (g, gs) for $g \in G$ and $s \in S$. This definition results in double edges for generators $s \in S$ of order two, as $g = gss$ in this case. This point of view is not useful in the context of Chap. 6. We hence stick with the definition of non-oriented Cayley graphs throughout this book.

The following example illustrates that Cayley graphs of groups are by no means unique and come in many different shapes and forms.

Example 2.26 Figure 2.2 shows two different Cayley graphs for the additive group $\mathbb{Z}$. The one on the left is the Cayley graph with respect to the generating set containing the single element 1. The graph on the right is the Cayley graph of the same group with respect to the generating set $\{2, 3\}$. There are infinitely many other Cayley graphs for this group.

Fig. 2.2 Two Cayley graphs for the additive group $\mathbb{Z}$: On the left $\Gamma(\mathbb{Z}, \{1\})$, on the right $\Gamma(\mathbb{Z}, \{2, 3\})$

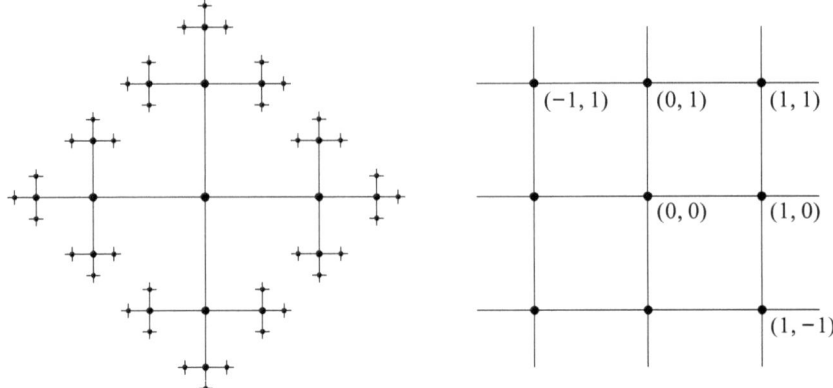

Fig. 2.3 The Cayley graph $\Gamma(F_2, \{a, b\})$ is the regular 4-valent tree (left) while $\Gamma(\mathbb{Z}^2, \{a, b\})$ is a regular square grid (right). Here $a = (1, 0)$ and $b = (0, 1)$

The free group on two elements has a regular 4-valent tree as its Cayley graph with respect to the standard generating set on two elements. This tree is pictured in Fig. 2.3 where we have omitted the labels of the edges.

As a third example consider the free abelian group on two generators. The standard generating set with two commuting elements produces a regular square grid as its Cayley graph, which is also shown in Fig. 2.3.

Finite groups obviously have finite Cayley graphs. The graph $\Gamma(\mathbb{Z}/\mathbb{Z}_6, \{\mathbb{1}\})$ is, for example, a hexagon while $\Gamma(\mathbb{Z}/\mathbb{Z}_6, \mathbb{Z}/\mathbb{Z}_6)$ is the complete graph on six vertices. Here a graph is a *complete graph* if every pair of its vertices in connected by an edge.

Cayley graphs have many nice properties. A few elementary ones follow directly from the definition.

Proposition 2.27 (Properties of Cayley Graphs) *The following properties hold for any finitely generated group G with finite generating set S.*

1. *$\Gamma(G, S)$ is connected.*
2. *$\Gamma(G, S)$ is regular and each vertex is contained in $|S \cup S^{-1}|$ edges.*
3. *$\Gamma(G, S)$ is locally finite.*

The Cayley graph allows us to interpret words in the generators as edge-paths between elements of the group. Recall that an edge-path in a graph is a sequence of vertices $(v_0, v_1, \ldots, v_k)$ such that any two subsequent vertices v_i and v_{i+1} are connected by an edge.

Definition 2.28 (Path Associated to a Word) Let G be a group finitely generated by S. Denote by Γ the Cayley graph of G with respect to S. Suppose $\bar{g} = s_1 s_2 \cdots s_k$ is a word in the generators $s_i \in S$ representing an element $g \in G$. Then for every

$x \in S$ the word $\bar{g}$ defines an edge-path γ_g in Γ connecting x with xg as follows:

$$\gamma_g := (x, xs_1, xs_1s_2, \ldots, xs_1s_2 \cdots s_k = xg).$$

One can consider the Cayley graph of a group as a metric space and define the so called *word metric* on Γ.

Definition 2.29 (Word Metric, Length) Let G be a finitely generated group and $\Gamma = \Gamma(G, S)$ its Cayley graph with respect to a finite generating set S. The distance $d_S(g, h)$ of two vertices g and h is defined to be the minimal number of edges of an edge-path connecting g and h in Γ. We call d_S the *word metric* on G with respect to S. The *length* of an element $g \in G$, denoted by $l(g)$, is the distance between g and the identity. More formally $l(g) := d_S(g, \mathbb{1})$.

It is not hard to check that $d_S : G \times G \to \mathbb{N}$ is in fact a metric on the set of vertices of Γ. This metric coincides with the distance between vertices as defined in Definition 2.6 and also with the distance between the vertices when viewed as points in the metric realization of Γ, compare Definition 2.6. The edge-path from x to xg obtained from a minimal word for the element g has length equal to the distance of x and xg measured in the word-metric.

For an infinite group G we have hence produced infinitely many (discrete) metric spaces – one for each generating system. Cayley graphs with respect to distinct generating systems are in general non-isomorphic. However, there is a way in which they (at least coarsely) agree.

Definition 2.30 (Quasi-Isometry)   A map $f : X \to Y$ between two metric spaces (X, d_X) and (Y, d_Y) is a *quasi-isometric embedding* if there exist constants C and D in $\mathbb{R}_{>0}$ such that for all $x, y \in X$

$$\frac{1}{C} d_X(x, y) - D \le d_Y(f(x), f(y)) \le C d_X(x, y) + D.$$

A map $g : X \to Y$ has *finite distance* from f if there exists a constant $c \in [0, \infty)$ such that for all $x \in X$ one has $d_Y(f(x), g(x)) \le c$. The map f is a *quasi-isometry* if it is a quasi-isometric embedding, which has a *quasi-inverse*. This is equivalent to the existence of a quasi-isometric embedding $g : Y \to X$ such that $g \circ f$ has finite distance from the identity map $\mathbb{1}_X$ on X and such that $f \circ g$ has finite distance from the identity map $\mathbb{1}_Y$ on Y.

Alternatively, a map f is a quasi-isometry if and only if it is a quasi-isometric embedding with *an R-dense image* in Y, that is, there exists a constant $R < \infty$ such that for all $y \in Y$ there exists $x \in X$ with $d_Y(f(x), y) < R$. Every isometry is also a quasi-isometry.

Example 2.31   The following maps are quasi-isometries:

1. $f : \mathbb{Z}^2 \hookrightarrow \mathbb{R}^2$ where we equip $\mathbb{Z}^2$ with the restricted Euclidean metric,
2. $f : \mathbb{R} \to \mathbb{Z} : x \mapsto \lfloor x \rfloor$, and

3. $f : X \to \{x\}$ for any non-empty, bounded metric space X and the space $\{x\}$ consisting of a single point.

The following maps are not quasi-isometries:

4. $f : \mathbb{R} \to \mathbb{R} : n \mapsto n^2$, and
5. any map from $(\mathbb{R}^2, d_{eucl})$ to the hyperbolic plane $\mathbb{H}^2$ with its standard metric. This can be seen as follows. Being hyperbolic is a quasi-isometry invariant. For a proof see for example [Lö17, Prop. 7.2.9]. While one can show that $\mathbb{H}^2$ is hyperbolic (see [BH99, Thm. I.1A.6]), the Euclidean plane is not. Hence the two spaces are not quasi-isometric.

Using Cayley graphs one can show that the following pairs of groups are quasi-isometric:

1. $\mathbb{Z}$ and $\mathbb{Z}/2\mathbb{Z} * \mathbb{Z}/2\mathbb{Z}$, and
2. any pair of finite groups.

It turns out that all Cayley graphs for a same group are quasi-isometric. A proof of this fact can be found on p. 138ff in [BH99].

Proposition 2.32 (Quasi-Isometric Cayley Graphs) *Let G be a finitely generated group. Let S and T be two finite generating sets for G. Then (G, d_S) and (G, d_T) are quasi-isometric. The identity map $\mathbb{1} : G \to G$ is one such quasi-isometry. Further (G, d_S) is quasi-isometric to the metric realization of the Cayley graph $\Gamma(G, S)$.*

A metric realization of the Cayley graph $\Gamma(G, S)$ is a connected space while the group G equipped with the metric d_S is a discrete space. The respective metrics are compatible in the sense that the distance between vertices g, h in $\Gamma(G, S)$ is exactly the same as the distance $d_S(g, h)$ between them in the (discrete) metric space (G, d_S). Both spaces are quasi-isometric.

Any finitely generated group naturally acts on its Cayley graphs. These actions have particularly nice properties characterized in Definition 2.34.

Example 2.33 A finitely generated group G always acts by left-multiplication on all of its Cayley graphs $\Gamma := \Gamma(G, S)$ as follows. First define for a fixed $g \in G$ the left-multiplication map $l_g : G \to G$ by putting $l_g(h) = gh$. In order to define an action of G on Γ we need to assign to every element in G an automorphism of the graph Γ, i.e. a bijection on the set of vertices that maps every pair of vertices connected by an edge to a pair of vertices connected by an edge and hence keeps adjacency of vertices intact. The maps l_g introduced above have this property as multiplication is done on the left while edges in Γ are defined by right-multiplication of generators. We may hence define the left-multiplication action of G on Γ by

$$G \to \mathrm{Aut}(\Gamma) : g \mapsto l_g.$$

One can show that this map is a well defined homomorphism. This action automatically preserves labels of edges and is geometric in the sense of Definition 2.34.

Definition 2.34 (Geometric Actions)   An action of a group G on a metric space (X, d) is *geometric* if the following conditions are satisfied

1. the action is properly discontinuous, that is, for all compact sets $K \subset X$ the set $\{g \in G \mid g.K \cap K \neq \emptyset\}$ is finite,
2. X/G is compact with respect to the quotient topology, and
3. G acts by isometries.

Remark 2.35   Recall that a space is *proper* if every closed ball $\overline{B}_r(x)$ is compact. Moreover a map is proper if all preimages of compact sets are compact. Hence, an action is properly discontinuous if and only if the map

$$G \times X \to X \times X : (g, x) \mapsto (x, g.x)$$

is proper when G carries the discrete topology.

Maybe the most fundamental theorem in geometric group theory is the following observation, which goes back to Švarc[1] [Shv55], see also [Mil68]. It states that any group, which acts nicely on a good metric space, is quasi-isometric to the space. More precisely, one can show:

Theorem 2.36 (Švarc–Milnor Theorem) *Suppose G acts geometrically on a nonempty, proper, geodesic metric space (X, d), then G is finitely generated and quasi-isometric to X.*

One may alternatively assume that X is a length space, which is slightly stronger than being proper and geodesic. Compare [BH99, I.8.4]. For a proof of the Švarc–Milnor theorem see [BH99, I.8.19] or [Lö17].

The main step in the proof of the Švarc-Milnor theorem is to write down a finite generating set. This is done as follows. Let G and X be as in Theorem 2.36 and let x_0 be a point in X. Define

$$A := \{g \in G \mid g.B_r(x_0) \cap B_r(x_0) \neq \emptyset\}$$

where the radius $r > 0$ is chosen large enough such that for a (fixed) compact set C with the property that $G.C = X$ one has $C \subset B_{r/3}(x_0)$. Then one can show that A is a generating set of G. The quasi-isometry from the Cayley graph of G with respect to A to the space (X, d) is given by the orbit map of the action. That is, the vertex g of $\Gamma(G, A)$ is mapped onto $g.x_0 \in X$.

---

[1] Note that also Schwarz, Shvarts and Svarc are used.

2.3   Exercises

Exercise 2.37 Consider the maximum norm on $\mathbb{R}^n$ given by

$$\|x\|_\infty := \max_i \{|x_i|, 1 = 1, \ldots, n\}$$

for all $x = (x_1, x_2, \ldots, x_n)^T \in \mathbb{R}^n$. Define the distance $d_\infty(x, y)$ of any pair of points $x, y \in \mathbb{R}^n$ by

$$d_\infty(x, y) := \|x - y\|_\infty.$$

Decide whether $(\mathbb{R}^n, d_\infty)$ is (locally) geodesic.

Exercise 2.38 Let V be a normed $\mathbb{R}$-vector space equipped with the metric induced by the norm. Prove the following statements.

(a) For any choice of $x, y \in V$ the map

$$\gamma : [0, \|y - x\|] \to V,$$

$$t \longmapsto ((\|y - x\| - t) \cdot x + t \cdot y) \cdot \frac{1}{\|y - x\|}$$

defines a geodesic from x to y. The space V is a geodesic metric space.

(b) V is uniquely geodesic if and only of the closed unit ball $\bar{B}_1(0) \subset V$ is strongly convex. Here the set $\bar{B}_1(0)$ is *strongly convex* if for all $u_1 \neq u_2 \in \mathbb{R}^n$ of norm 1 the inequality $\|(1 - t)u_1 + tu_2\| < 1$ is satisfied for all $t \in (0, 1)$.

Exercise 2.39 Let (X, d) be a complete metric space and suppose that for all $x, y \in X$ there exists a *midpoint*, that is, a point $m \in X$ such that

$$d(x, m) = \frac{1}{2} d(x, y) = d(m, y).$$

Show that X is a geodesic space.

Exercise 2.40 The product (X, d) of two metric spaces (X_1, d_1) and (X_2, d_2) is the set $X = X_1 \times X_2$ equipped with the *product metric* defined by

$$d((x_1, x_2), (y_1, y_2)) := \sqrt{d_1(x_1, y_1)^2 + d_2(x_2, y_2)^2}.$$

Prove that

1. the map d from above is indeed a metric on X.
2. the space X is complete if and only if both X_1 and X_2 are complete.

Exercise 2.41 (Alexandrov's Lemma) Let a, b, b' and c be four distinct points in $\mathbb{R}^2$ such that b and b' are on different sides of the line spanned by a, c. Consider the triangle $\Delta = \Delta(a, b, c)$ with inner angles α, β, γ at a, b, and c, respectively. And let a triangle $\Delta' = \Delta(a, b', c)$ (sharing two vertices with Δ) with inner angles α', β', γ' at a, b' and c be given, such that $\gamma + \gamma' \geq \pi$. Then show

$$d(b, c) + d(b'c) \leq d(b, a) + d(b', a).$$

Exercise 2.42 (Alexandrov's Lemma, Continued) With the setup as in Exercise 2.41 let $\bar{\Delta} = \Delta(\bar{a}, \bar{b}, \bar{b}')$ be a Euclidean triangle defined by the three conditions

$$d(\bar{a}, \bar{b}) = d(a, b), \ d(\bar{a}, \bar{b}') = d(a, b') \ \text{and} \ d(\bar{b}, \bar{b}') = d(b, c) + d(c, b').$$

Denote its angles by $\bar{\alpha}, \bar{\beta}, \bar{\beta}'$. Further let $\bar{c} \in \text{seg}(\bar{b}, \bar{b}')]$ be the point in $\mathbb{R}^2$ determined by $d(\bar{b}, \bar{c}) = d(b, c)$. Prove that

1. $\bar{\alpha} \geq \alpha + \alpha'$,
2. $\bar{\beta} \geq \beta$,
3. $\bar{\beta}' \geq \beta'$, and
4. $d(\bar{a}, \bar{c}) \geq d(a, c)$.

Show in addition that if one of the inequalities in items 1 to 4 is an equality, then also the others are. Show that this happens if and only if $\gamma + \gamma' = \pi$.

The statement in the previous exercise holds true (with minor modifications) in all model spaces.

Exercise 2.43 Let T_k be the $2k$-regular tree. Show that for all $k, k' \geq 2$ the tree T_k is quasi-isometric to the tree $T_{k'}$.

Chapter 3
Non-positive Curvature

Groups acting on non-positively curved spaces satisfy many interesting properties. One widely used concept of non-positive curvature for metric spaces is the notion of a CAT(0) space. The CAT(0) property is a thinness condition on triangles which requires them to be at most as thick as those in Euclidean space. The main purpose of this chapter is to introduce the notion of a CAT(0) space and to prove some of their properties, such as the existence of projections onto convex subsets or the Bruhat-Tits fixed point theorem.

More on the geometry of non-positively curved metric spaces can be found in the little green book by Ballmann [Bal95], the book by Burago, Burago and Ivanov [BBI01] or in Bridson and Haefliger's monograph [BH99].

3.1 Metric Spaces of Non-positive Curvature

As mentioned above, the CAT(0) condition compares the diameter of triangles in a metric space with the diameter of triangles with the same side lengths in Euclidean space. More generally, the CAT(κ) property captures the thickness of triangles in a given geodesic metric space in comparison with triangles of the same side lengths in the model spaces M_κ^2 of constant sectional curvature κ.

One says that a metric space has curvature less than or equal to κ if all of its triangles are thinner than the respective comparison triangle in the model space M_κ^2. Similarly, one speaks of non-positive curvature if all triangles are thinner than the comparison triangles in the Euclidean plane.

In this section we introduce the model spaces in question, define CAT(0) spaces and prove the Bruhat–Tits fixed point theorem.

P. Schwer, *CAT(0) Cube Complexes*, Lecture Notes in Mathematics 2324,
https://doi.org/10.1007/978-3-031-43622-2_3

3.1.1   Model Spaces of Constant Sectional Curvature

In this subsection we introduce the three classes of model spaces with constant
sectional curvature κ. They serve as reference spaces for defining the curvature of
an arbitrary metric space. It suffices to consider the following three spaces in order
to understand the material contained in this book. For a detailed account on model
spaces of constant sectional curvature κ see, for example, Chapter I.2 and I.6 of
[BH99].

Definition 3.1 (Model Spaces) The 2-dimensional model spaces M_κ^2 with $\kappa \in$
$\{-1, 0, 1\}$ are defined follows:

1. in case $\kappa = 1$ the space $M_1^2 = \mathbb{S}^2$ is the 2-dimensional unit sphere together with
   the round metric. That is, $\mathbb{S}^2 \cong S_1(0) \subset \mathbb{R}^3$. The diameter of this space, i.e. the
   maximal distance between a pair of points in $\mathbb{S}^2$, equals $D_1 = \pi$.
2. in case $\kappa = 0$ the space $M_0^2 = \mathbb{E}^2$ is the plane $\mathbb{R}^2$ equipped with the standard
   Euclidean metric. Its diameter is infinite. Hence, put $D_0 = \infty$.
3. in case $\kappa = -1$ the space $M_{-1}^2 = \mathbb{H}^2$ is the hyperbolic plane. The diameter D_{-1}
   of this space is also ∞.

All other model spaces M_κ^2 with $\kappa \in \mathbb{R}$ are obtained from M_1^2 or M_{-1}^2 by
appropriately rescaling the metric. For $\kappa > 0$ rescale the metric on the round sphere
M_1^2 by $\frac{1}{\sqrt{\kappa}}$ to obtain M_κ^2. If $\kappa < 0$ then M_κ^2 is obtained from $\mathbb{H}^2$ by multiplying the
metric by $\frac{1}{\sqrt{-\kappa}}$. Illustrations of the three main model spaces with $\kappa \in \{-1, 0, 1\}$ are
provided in Fig. 3.1.

One can prove the following characterization.

Theorem 3.2 (Characterization of Model Spaces) *For $\kappa \in \mathbb{R}$ the model space
M_κ^2 is the unique complete, simply connected Riemannian manifold of dimension
two with constant sectional curvature κ. We write D_κ for the diameter of M_κ^2.*

Model spaces also exist in higher dimensions. We learn about two classes in
Definition 4.1.

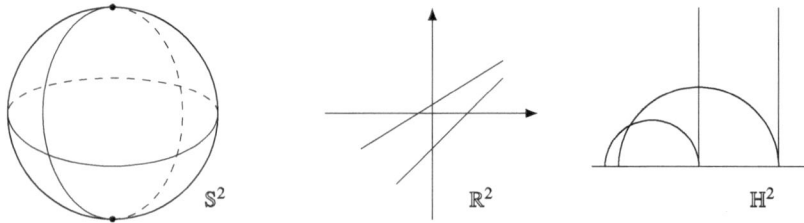

Fig. 3.1 From left to right: the model spaces M_κ^2 for $\kappa = 1, 0$ and -1 showing some of their
geodesics

3.1.2 Definition of CAT(0) Spaces

The goal of this sub-section is to define CAT(0) spaces. In order to do so we first need an appropriate notion of triangles and their CAT(0) property.

Definition 3.3 (Geodesic Triangles)   Let x, y and z be a triple of points in a metric space (X, d). A *(geodesic) triangle* $\Delta = \Delta(x, y, z)$ on the points x, y, z is the union of three geodesic segments $\text{seg}_{\gamma_1}(x, y)$, $\text{seg}_{\gamma_2}(y, z)$ and $\text{seg}_{\gamma_3}(z, x)$. Refer to these segments as the *sides* of Δ.

In general, there might be more than one geodesic triangle on the same three vertices x, y, z. This is always the case when there is more than one geodesic connecting any two of the points. In an arbitrary metric space not every triple of points spans a geodesic triangle as it can happen that pairs of points are not connected by any geodesic.

Definition 3.4 (Comparison Triangles)   Let (X, d) be a metric space, let x, y and z be points in X and let $\Delta(x, y, z)$ be a triangle. A *comparison triangle* of $\Delta(x, y, z)$ in the 2-dimensional model space (M_κ^2, d_κ) is a triangle $\Delta(\bar{x}, \bar{y}, \bar{z})$ with vertices $\bar{x}, \bar{y}, \bar{z}$ in M_κ^2 such that $d_\kappa(\bar{x}, \bar{y}) = d(x, y)$, $d_\kappa(\bar{y}, \bar{z}) = d(y, z)$ and $d_\kappa(\bar{x}, \bar{z}) = d(x, z)$.

The comparison triangle has the same side lengths. A first exercise is proving such a comparison triangle always exist and is unique up to congruence, see Exercise 3.38. Using comparison triangles one can define curvature bounds for geodesic metric spaces.

Geodesics and geodesic segments in the Euclidean plane are unique. We may thus write $\text{seg}(x, y)$ instead of $\text{seg}_\gamma(x, y)$ for segments in $\mathbb{R}^2$.

Definition 3.5 (CAT(κ) Property)   Let (X, d) be a metric space and $\Delta = \Delta(a, b, c)$ a triangle in X. Let $\bar{\Delta}$ be its comparison triangle in M_κ^2. Fix a point p on the side $\text{seg}_\gamma(a, b)$ of Δ. A *comparison point* $\bar{p}$ of p is a point on the side $\text{seg}(\bar{a}, \bar{b})$ of $\bar{\Delta}$ such that $d_\kappa(\bar{p}, \bar{a}) = d(p, a)$ and $d_\kappa(\bar{p}, \bar{b}) = d(p, b)$.

A triangle Δ in X *satisfies the* CAT(κ) *property* if

1. its circumference is less than $2D_\kappa$ and if
2. for all points p, q on the sides of Δ and their respective comparison points $\bar{p}, \bar{q}$ in a comparison triangle $\bar{\Delta}$ one has

$$d = d(p, q) \le d_\kappa(\bar{p}, \bar{q}) = \bar{d}.$$

Figure 3.2 shows a triangle Δ in a metric space (on the left) and its comparison triangle in the Euclidean plane (on the right). If for all p and q in Δ the distance $\bar{d}$ of their comparison points $\bar{p}$ and $\bar{q}$ in the comparison triangle is at least as big as the distance d of p and q in Δ then Δ satisfies the CAT(0) condition.

Spaces are CAT(κ) if all their triangles satisfy the CAT(κ) property.

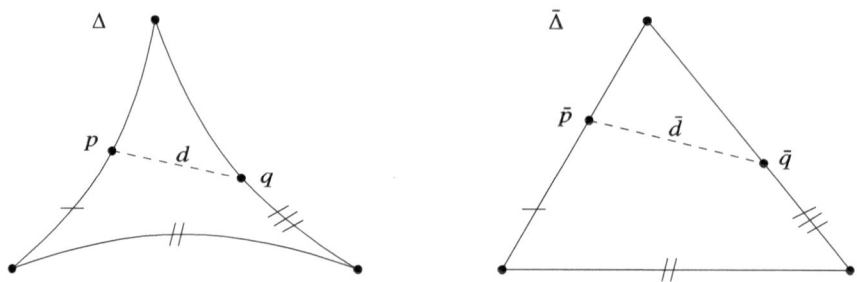

Fig. 3.2 The CAT(0) property for triangles. Here d is the distance $d_X(p, q)$ and $\bar{d} = d_\kappa(\bar{p}, \bar{q})$

Definition 3.6 (CAT(κ) **Spaces**) A CAT(κ) *space* is a D_κ-uniquely geodesic metric space (X, d) in which all triangles satisfy the CAT(κ) property, in the sense just defined. We say that (X, d) is *locally* CAT(κ) if for all $x \in X$ there exists $r_x > 0$ such that $B_{r_x}(x)$ is a CAT(κ) space with respect to the restricted metric.

According to most sources the acronym CAT was coined by Misha Gromov and stands for Cartan, Alexandrov and Toponogov who where among the first describing spaces of this kind. Locally CAT(κ) spaces are sometimes called *non-positively curved* and CAT(κ) spaces are also known as Alexandrov spaces with upper curvature bound.

Example 3.7 Some examples of CAT(κ) spaces:

1. n-dimensional Euclidean space is CAT(0) for all n.
2. Metric graphs without circuits are CAT(0). For details on metric graphs see Definition 2.6.
3. $\mathbb{R}$-trees are CAT(0). Compare Exercise 3.44.
4. Simply connected Riemannian manifolds of constant sectional curvature κ are CAT(κ). Spheres are, for example, CAT(1) and hyperbolic space is an example of a CAT(-1) space. See [BH99, Thm. II.1A.6] for further details.

We now consider some elementary properties of CAT(κ) spaces.

Proposition 3.8 (Properties of CAT(κ) **Spaces)** *If (X, d) is a CAT(κ) space then*

1. *X is D_κ-uniquely geodesic.*
2. *the ball B_r of radius r is contractible for all $r < D_\kappa$. In particular X is contractible if $\kappa \leq 0$.*
3. *every local geodesic of length $< D_\kappa$ is a geodesic.*

Proof To prove item 1, let $x, y \in X$ be two points at distance $d(x, y) < D_\kappa$. Further let γ and γ' be two distinct geodesics from x to y as illustrated in Fig. 3.3. Then

$$\mathrm{seg}_\gamma(x, y) = \gamma([0, d(x, y)]) \neq \gamma'([0, d(x, y)]) = \mathrm{seg}_{\gamma'}(x, y).$$

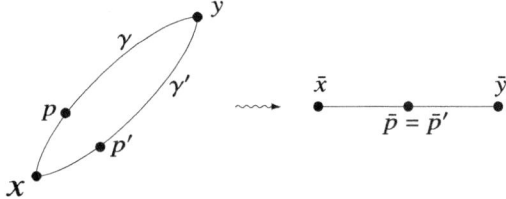

Fig. 3.3 Non-unique geodesics yield degenerate comparison triangles

Choose points $p \in \text{seg}_\gamma(x, y)$ and $p' \in \text{seg}_{\gamma'}(x, y)$ such that $d(x, p) = d(x, p')$. The comparison triangle $\bar{\Delta}$ for $\Delta = \Delta(x, p, y)$ with sides

$$\gamma([0, d(x, p)]), \; \gamma([d(x, p), d(x, y)]) \text{ and } \text{seg}_{\gamma'}(x, y)$$

is degenerate and hence $\bar{p} = \bar{p}'$. From the CAT(κ) property we may then deduce that $0 = d(\bar{p}, \bar{p}') \geq d(p, p') \geq 0$ and hence $p = p'$.

Item 2 can be deduced from item 1 as follows: For any radius $0 < r < D_\kappa$ the map $f : B_r(x) \times [0, 1] \to X$ that sends pairs (y, t) to the point p on $\text{seg}_\gamma(x, y)$ with $d(y, p) = t d(x, y)$ is a continuous retraction from the ball $B_r(x)$ to x, since by item 1 there is a unique geodesic connecting x and y if $d(x, y) < D_\kappa$. Thus, item 2.

Suppose, for some $l < D_\kappa$, that $\gamma : [0, l] \to X$ is a local geodesic and consider the set $S := \{t \mid \gamma|_{[0,t]} \text{ is a geodesic}\}$. Since S is closed in $[0, l]$ it remains to show that it is also open in $[0, l]$. By definition of local geodesics there is $0 < t_0 < l$ and $\varepsilon > 0$ such that γ restricted to $[t_0 - \varepsilon, t_0 + \varepsilon]$ is a geodesic. Consider the triangle $\Delta = \Delta(\gamma(0), \gamma(t_0), \gamma(t_0 + \varepsilon))$ where the sides are $\gamma([0, t_0]), \gamma([t_0, t_0 + \varepsilon])$ and the unique geodesic σ from $\gamma(0)$ to $\gamma(t_0 + \varepsilon)$. One can prove that the comparison triangle $\bar{\Delta}$ is degenerate: Suppose it was not degenerate and apply the CAT(κ) condition to points $x \in \text{seg}_\gamma(\gamma(0), \gamma(t_0))$ and $y \in \text{seg}_\gamma(\gamma(t_0), \gamma(t_0 + \varepsilon))$ where both x and y are chosen to lie closer than ε to $\gamma(t_0)$. This yields a contradiction to the fact that $\gamma|_{[t_0 - \varepsilon, t_0 + \varepsilon]}$ is a geodesic. Thus, $\bar{\Delta}$ is in fact degenerate which allows us to conclude that $\gamma|_{[0, t_0 + \varepsilon]}$ is a geodesic. Moreover $d(\gamma(0), \gamma(t_0 + \varepsilon)) = t_0 + \varepsilon$ and $(t_0, t_0 + \varepsilon) \subset S$. Therefore, S is open.   □

3.1.3   *Properties of* CAT(0) *Spaces*

This subsection contains a few basic properties of CAT(0) spaces. We introduce convex subsets and projections onto them and note that new CAT(0) spaces can be constructed from known ones by taking convex subsets (see Exercise 3.42). We also prove the Bruhat–Tits theorem on centers of bounded subsets in CAT(0) spaces, two fixed point theorems and a local to global principle of isometries.

Definition 3.9 (Convex Sets)   A subset C of a metric space (X, d) is *convex* if for all $x, y \in C$ there exists a geodesic $\gamma : x \rightsquigarrow y$ connecting x and y and all geodesic segments $\text{seg}(x, y)$ are contained in C.

Example 3.10   Convex subsets of the n-dimensional Euclidean space are CAT(0) with respect to the restricted metric. This is a special case of the general fact that convex subsets of CAT(0) spaces are again CAT(0). Compare Exercise 3.42.

The projection map introduced in the next proposition is an important and powerful tool when working with CAT(0) spaces.

Proposition 3.11 (Projections onto Convex Sets) *Let X be a complete* CAT(0) *space and $\emptyset \neq A \subset X$ a closed and convex subset. Then*

1. for all $x \in X$ there exists a unique point $\pi_A(x) \in A$ such that

$$d(x, \pi_A(x)) = \inf_{a \in A} d(x, a).$$

2. $\pi_A : X \to A : x \mapsto \pi_A(x)$ is distance non-increasing, that is,

$$d(\pi_A(x), \pi_A(y)) \leq d(x, y) \text{ for all } x, y \in X.$$

3. if $y \in \text{seg}(x, \pi_A(x))$ then $\pi_A(x) = \pi_A(y)$.

Proof We follow the proof provided in [BH99, Prop II.2.4]. To show item 1 fix $x \in X$ and let $(y_n)_n$ be a sequence in A such that the distance $d(x, y_n)$ converges to $d(x, A)$ for large n. We show that $(y_n)_n$ is a Cauchy sequence. Consider a comparison triangle $\bar{\Delta}$ on $\bar{x}, \bar{y}_m, \bar{y}_m$ for the triangle spanned by x, y_m and y_n for some fixed m and n. Extend the triangle $\bar{\Delta}$ by a fourth vertex $\bar{p}$ to a parallelogram in the plane, where p is chosen such that $d(\bar{p}, \bar{y}_k) = d(\bar{x}, \bar{y}_k)$ with $k \in \{m, n\}$.

By the parallelogram law of Euclidean geometry one may then conclude that

$$2(d^2(\bar{x}, \bar{y}_n) + d^2(\bar{x}, \bar{y}_m)) = d^2(\bar{x}, \bar{p}) + d^2(\bar{y}_m, \bar{y}_n).$$

Put $\delta := d(x, A)$ and let ε be small compared to δ. Choose m and n large enough so that both $d(x, y_n)$ and $d(x, y_m)$ are strictly less than $\delta + \varepsilon$. By hypothesis there exists N such that all $m, n > N$ have this property.

By the CAT(0) property of X one concludes that for the midpoint m of y_m and y_n and its comparison point $\bar{m}$ one has

$$d(\bar{x}, \bar{m}) \geq d(x, m) \geq d(x, A).$$

Combining the fact that $d(x, y_n), d(x, y_m) < \delta + \varepsilon$ with the last inequality we may conclude

$$d^2(y_n, y_m) \le d^2(\bar{y}_n, \bar{y}_m)$$
$$= 2(d^2(\bar{x}, \bar{y}) + d^2(\bar{x}, \bar{y}_m)) - d^2(\bar{x}, \bar{p})$$
$$\le 4(d(x, A) + \varepsilon)^2 - 4d^2(x, A)$$
$$= 8\varepsilon d(x, A) + 4\varepsilon^2$$
$$= 4(2\varepsilon d(x, A) + \varepsilon^2).$$

Hence, $(y_n)_n$ is a Cauchy sequence and has a limit point in C with the property that $\lim_{n\to\infty} d(x, y_n) = d(x, A)$. We put $\pi_A(x)$ to be the limit of $(y_n)_n$. Uniqueness then follows by construction.

To verify item 3 suppose that $\pi_C(y) \ne \pi_C(x)$. In this case one in particular has $d(y, \pi_C(y)) < d(y, \pi_C(x))$ and hence

$$d(x, \pi_C(y)) \le d(x, y) + d(y, \pi_C(y))$$
$$< d(x, y) + d(y, \pi_C(x)) = d(x, \pi_C(x))$$

where the last equality follows from the fact that $y \in \mathrm{seg}(\pi_C(x), x)$. But, this contradicts the minimality of the choice of $\pi_C(x)$.

The proof of item 2 is left as an exercise to the reader.   □

One can show that every finite set of points $x_1, \ldots, x_n$ in a non-positively curved, simply connected, complete Riemannian manifold M has a center. That is, the map that assigns to a point $x \in M$ the value $\sum_{i=1}^n d(x, x_i)^2$ attains a minimum. This implies that a compact group of isometries of M always admits a fixed point. This statement was generalized by Bruhat and Tits [BT84] to Euclidean buildings, which are a special class of CAT(0) spaces. Let us have a look at the analogous statement for general CAT(0) spaces.

Definition 3.12 (Radius) The *radius* of a bounded set Y in a metric space (X, d) is given by

$$\mathrm{rad}(Y) := \inf\{r \in \mathbb{R}_{>0} \mid Y \subset \bar{B}_r(x) \text{ for some } x \in X\}.$$

Theorem 3.13 (Bruhat–Tits Theorem) *Let $X \ne \emptyset$ be a complete CAT(0) space and let $Y \subset X$ be a bounded subset. Then there exists a unique point $p \in X$, called the center of Y, such that $\bar{B}_r(p) \supset Y$ with $r = \mathrm{rad}(Y)$.*

Proof Suppose there are two points q, q' in X such that $\bar{B}_{r+\varepsilon}(q) \cap \bar{B}_{r+\varepsilon}(q') \supset Y$. Then one can deduce from the parallelogram law (similar to what we have done in the proof of Proposition 3.11) that

$$d(q, q') \leq 4(r^2 + 2r\varepsilon + \varepsilon^2) - 4r^2.$$

This implies uniqueness of the center as follows: suppose now that q and q' are both centers of Y. Then we can choose $\varepsilon = 0$ in the inequality above and deduce that $d(q, q') = 0$ and hence $q = q'$.

To show existence let (p_n) be a sequence of points in X such that $\overline{B_{r+\frac{1}{n}}(p_n)} \supset Y$. But then, the inequality stated above yields that $(p_n)_n$ is a Cauchy sequence, which has a limit point p in X. By construction we have for all $y \in Y$ that $\lim_{n \to \infty} d(y, p_n) \leq r$ and thus $\bar{B}_r(p) \supset Y$. Hence, the assertion.   □

Note that there are examples subsets Y in spaces X for which the center of Y obtained from the Bruhat–Tits theorem is not contained in Y.

We now shift our attention to fixed point sets of isometries of $\mathrm{CAT}(0)$ spaces and prove that they are well behaved subsets of those spaces.

Proposition 3.14 (The Little Fixed Point Theorem) *The fixed point set of an isometry of a* $\mathrm{CAT}(0)$ *space is closed and convex.*

Proof First note that fixed point sets of continuous self-maps on Hausdorff-spaces are closed. Thus, $\mathrm{Fix}(\phi)$, where $\phi : X \to X$ is an isometry of a $\mathrm{CAT}(0)$ space, needs to be closed. In order to see that the fixed point set is convex it is enough to prove that geodesics between fixed points x, y are pointwise fixed. But, this may be deduced from the fact that $\mathrm{CAT}(0)$ spaces are uniquely geodesic, see item 1 of Proposition 3.8.   □

Considering group actions one can prove stronger fixed point theorems.

Theorem 3.15 (Bruhat–Tits Fixed Point Theorem) *Suppose a group* G *acts isometrically on a complete* $\mathrm{CAT}(0)$ *space* $X \neq \emptyset$. *Then the following are equivalent.*

1. *Every G-orbit in X is bounded.*
2. *G has a bounded orbit in X.*
3. *The fixed point set $X^G = \{x \in X \mid g(x) = x \text{ for all } g \in G\}$ is nonempty and convex.*

Proof It is clear that the first item implies the second.

To see that 2 implies 3 argue as follows. Choose a point x_0 in X such that its orbit $Y := G.x_0$ under G is bounded. In case G is finite each orbit is bounded and contained in a ball of radius $\max\{d(g.x, g'.x) \mid g, g' \in G\}$. From the Bruhat–Tits theorem 3.13 we obtain a unique center p of Y. Since the set Y is invariant under the G-action its center is also invariant under the action and hence fixed by G. Thus $X^G \neq \emptyset$. Convexity of the fixed point set follows from the fact that X is uniquely geodesic. Each geodesic connecting a pair of fixed points has to be mapped isometrically onto itself by an isometry. It is hence pointwise fixed.

It remains to prove that item 3 implies item 1. Choose an element x of the fixpoint set X^G and let y be an arbitrary element of X. As the action preserves distances and x is fixed, the orbit $G.y$ is contained in the ball $B_{d(x,y)}(x)$ of radius $d(x, y)$ around x. Therefore $G.y$ is a bounded set. Since y was chosen arbitrarily item 1 follows.

$\square$

CAT(0) spaces have very rigid local-to-global properties, which we now high-light.

Definition 3.16 (Local Isometries)   Let (X, d_X) and (Y, d_Y) be metric spaces. A map $f : X \to Y$ is a *local isometric embedding*, or short *local isometry*, if for every $x \in X$ there exists an $\varepsilon > 0$ such that $f|_{B_\varepsilon(x)}$ is an isometry onto its image.

Proposition 3.17 (From Local to Global Isometries) *Let X be a complete geodesic metric space and Y a CAT(0) space. Every local isometry $g : X \to Y$ from X to Y is an isometric embedding, that is, an isometry from X to $g(X) \subset Y$.*

Proof Let $g : X \to Y$ be a local isometry and $\gamma : [a, b] \to X$ a geodesic from x to y. Then $g \circ \gamma$ is a local geodesic in Y. Since Y is uniquely geodesic Proposition 3.8, item 3 implies that $g \circ \gamma$ is a geodesic and

$$d_Y(g(x), g(y)) = l(g \circ \gamma) = l(\gamma) = d_X(x, y)$$

where the second equality holds since the lengths of a curve is invariant under local isometries. $\square$

Theorem 3.18 contains (without proof) another fundamental property of CAT(0) spaces. This local-to-global principal is a natural analog to a similar statement in differential geometry. It allows to transfer local properties of a space X to global properties of its universal cover. For a proof see II.4.1 on page p. 193 in Bridson and Haefliger's book [BH99] or Chapter 1.4 in Ballmann's lecture notes [Bal95].

Theorem 3.18 (Cartan–Hadamard Theorem) *Let (X, d) be a complete, connected metric space. Then the following hold:*

1. *If d is locally convex, then the associated length metric on the universal cover $\tilde{X}$ of X is (globally) convex and $\tilde{X}$ is uniquely geodesic.*
2. *If (X, d) is locally CAT(0), then $\tilde{X}$ is (globally) CAT(0) with respect to the length metric induced by d.*

This directly implies the following corollary.

Corollary 3.19 *Every simply connected, complete, locally CAT(0) metric space is CAT(0). In particular universal covers of locally CAT(0) spaces are CAT(0).*

3.2   Angles in CAT(0) Spaces

In general, it is hard to test whether a given space is CAT(0). A characterization due
to Alexandrov uses angles to describe CAT(0) spaces. In this subsection we exhibit
the local structure of a CAT(0) space and introduce angles in this setting. We finish
by stating Alexandrov's criterion in Proposition 3.27.

Definition 3.20 (Angles in CAT(0) Spaces)   Let X be a CAT(0) space and let
$\gamma, \eta : [o, \varepsilon] \to X$ be two nontrivial geodesics with $\gamma(0) = \eta(0) = p$. We define
the *angle* $\alpha = \angle_p(\gamma, \eta)$ between η and γ at p by the following formula:

$$\sin\left(\frac{\alpha}{2}\right) = \lim_{t \to 0} \frac{d(\gamma(t), \eta(t))}{2t}.$$

Lemma 3.21   *The angle is a well defined notion.*

Proof The metric of every CAT(0) space is a convex function. By Definition 3.45
we have hence for all $0 < s < t < \varepsilon$ that

$$0 \le \frac{d(\gamma(s), \eta(s))}{2s} \le \frac{d(\gamma(t), \eta(t))}{2t} \le 1.$$

This implies that the limit in Definition 3.20 exists and is at most 1. Thus, the angle
is well defined.   □

The angle defines a pseudo metric on the set of all paths that start in a fixed
basepoint p. We prove this fact as Theorem 3.25 and prepare us with the following
notation and two technical lemma.

Notation 3.22   Let X be a CAT(0) space and fix a point p in X. For any pair of
points $x, y \in X \setminus \{p\}$ there exist unique geodesics $\gamma : p \rightsquigarrow x$ and $\eta : p \rightsquigarrow y$. We
may thus define the angle between x, y via

$$\angle_p(x, y) = \angle_p(\gamma, \eta).$$

In case $\Delta(\bar{p}, \bar{x}, \bar{y})$ is a comparison triangle for $\Delta(p, x, y)$ we may put

$$\overline{\angle}_p(x, y) := \angle_{\bar{p}}(\bar{x}, \bar{y})$$

and call it the *comparison angle of x and y at p*.

From Exercise 3.46 we have that the metric of a CAT(0) space is convex, which
easily implies that $\overline{\angle}_p(x, y) \ge \angle_p(x, y)$.

Lemma 3.23   *Let p, x and y be three distinct points in a CAT(0) space X. Let
further $\Delta(\bar{p}, \bar{x}, \bar{y})$ be the comparison triangle of the unique geodesic triangle on
p, x and y. Choose a point $\tilde{x}$ in $\mathbb{R}^2$ such that $d_{eucl}(\tilde{x}, \bar{p}) = d(x, p)$ and such that*

$\angle_{\bar{p}}(\tilde{x}, \tilde{y}) = \angle_p(x, y)$. *Then*

$$d(x, y) \geq d_{eucl}(\tilde{x}, \tilde{y}).$$

Proof This follows from the fact that angles in comparison triangles are at least as big as the corresponding ones in X itself, compare Notation 3.22. Hence, making the angle at $\bar{p}$ smaller while keeping two sides of the triangle the same length forces the third side (which has vertices $\tilde{x}$ and $\tilde{y}$) to get shorter. $\quad\square$

Lemma 3.24 *Let X be* CAT(0) *and let p, x and y be three distinct points in X. Let further $\gamma : p \rightsquigarrow x$ and $\eta : p \rightsquigarrow y$ be two nontrivial geodesics. Then for all $s, t > 0$ such that the geodesic are defined on s, respectively t, one has*

$$\angle_p(\gamma, \eta) \leq \overline{\angle}_p(\gamma(s), \eta(t)) \leq \angle_p(x, y).$$

Proof The first inequality follows from the fact that the metric is convex. To prove the second inequality proceed as follows. Consider the triangle $\Delta(p, x, y)$ and the points $\gamma(s), \eta(t)$ on its sides. We then construct two comparison triangles. One for $\Delta(p, x, y)$ and one for $\Delta(p, \gamma(s), \eta(t))$. The CAT(0) property of the triangle $\Delta(p, x, y)$ implies that

$$d(\gamma(s), \eta(t)) \leq d_{eucl}(\overline{\gamma(s)}, \overline{\eta(t)}).$$

Hence $\overline{\angle}_p(\gamma(s), \eta(t)) \leq \overline{\angle}_p(x, y)$. $\quad\square$

Now we are ready to prove the following theorem.

Theorem 3.25 (Angle Function Is a Pseudo-Metric) *For every* CAT(0) *space X the angle $\angle_p$ is a pseudo metric on $X \setminus \{p\}$ (or, respectively, on the set of all nontrivial geodesics starting in p) for all $p \in X$.*

Proof It is clear that the angle function is symmetric. We prove by contradiction that the triangle inequality holds. Suppose for a contradiction that there exist geodesics $\gamma, \gamma', \gamma''$ from p to x, y, z, respectively, such that

$$\angle_p(\gamma', \gamma'') > \angle_p(\gamma, \gamma') + \angle_p(\gamma', \gamma'').$$

Then there exists $\rho > 0$ such that

$$\angle_p(\gamma', \gamma'') > \angle_p(\gamma, \gamma') + \angle_p(\gamma', \gamma'') + 3\rho.$$

And there exists, by Lemma 3.24, a constant $\varepsilon > 0$ such that for all $0 < s, t \leq \varepsilon$ one has

$$\overline{\angle}_p(\gamma(s), \gamma'(t)) < \angle_p(\gamma, \gamma') + \rho.$$

and

$$\overline{\angle}_p(\gamma'(s), \gamma''(t)) < \angle_p(\gamma', \gamma'') + \rho.$$

Choose now points $\bar{p}, a, a''$ in $\mathbb{R}^2_{eucl}$ such that the following hold:

$$d(\bar{p}, a) = \varepsilon = d(\bar{p}, a'')$$

and such that the angle $\angle_{\bar{p}}(a, a'') = \tilde{\xi}$ satisfies

$$\pi > \tilde{\xi} > \angle_p(\gamma, \gamma'') - \rho.$$

Choose now a' on the line between a and a'' such that

$$\alpha := \angle_{\bar{p}}(a, a') > \angle_p(\gamma, \gamma') + \rho$$

and such that

$$\beta := \angle_{\bar{p}}(a', a'') > \angle_p(\gamma', \gamma'') + \rho.$$

Put $s := d(\bar{p}, a')$. Then the angle $\overline{\angle}_p(\gamma(\varepsilon), \gamma'(s))$ is strictly smaller than the sum $\angle_p(\gamma, \gamma') + \rho$. And hence $d(\gamma(\varepsilon), \gamma'(s)) < d_{eucl}(a, a')$. One also has that $d(\gamma'(s), \gamma''(\varepsilon)) < d_{eucl}(a', a'')$ and therefore

$$d(\gamma(\varepsilon), \gamma''(\varepsilon)) < d_{eucl}(a, a') + d_{eucl}(a', a'') = d(a, a'').$$

On the other hand $\overline{\angle}_p(\gamma(\varepsilon), \gamma''(\varepsilon)) > \tilde{\xi}$ and hence

$$d(\gamma(\varepsilon), \gamma''(\varepsilon)) > d_{eucl}(a, a'')$$

and we arrive at a contradiction.   □

Angles can be used to define a space of directions at a given point. This space plays the same role as the tangent space used in a smooth setting.

Definition 3.26 (Space of Directions) Let X be a CAT(0) space. We say that two nontrivial geodesics γ and η with $\gamma(0) = \eta(0) = p$ have *the same direction at* p if $\angle_p(\gamma, \eta) = 0$. This notion induces an equivalence relation on the set of geodesics starting at p and hence yields a quotient space called the *space of directions* at p. We denote this space by $\Sigma_p X$.

One can also use angles to characterize the CAT(0)-property in a geodesic metric space. We summarize this in the next proposition, which is sometimes referred to as Alexandrov's Lemma in the literature.

Proposition 3.27 (Alexandrov's Characterization of CAT(0) **Spaces)** *For a given geodesic metric space X the following are equivalent.*

1. *X is* CAT(0).
2. *For all geodesic triangles* $\Delta = \Delta(x, y, z)$ *in X and all* $p \in \text{seg}(x, y)$ *the comparison points* $\bar{p}$ *and* $\bar{z}$ *in* $\bar{\Delta}$ *satisfy* $d(\bar{p}, \bar{z}) \geq d(p, z)$.
3. *The* Alexandrov angle

$$\angle^A(\gamma, \gamma') := \lim_{t,t' \to 0} \sup \overline{\angle}(\gamma(t), \gamma'(t))$$

between two geodesic sides γ, γ' *of a geodesic triangle in X is at most the size of the corresponding angle in the comparison triangle. Here* $\overline{\angle}(\gamma(t), \gamma'(t))$ *is the angle in the comparison triangle on the vertices* $\gamma(0), \gamma(t), \gamma'(t)$.

A proof of this proposition can be found in [BH99, Thm. I.1.7]. The Alexandrov angle appearing in the last item of this proposition coincides with the definition of an angle given above. Compare also [BH99, I.1.12] for more on Alexandrov angles.

3.3 Flat Cones

The flat cone construction is a metrized cone over a metric space X which yields a CAT(0) space when taking the cone over a CAT(1) space, see Berestovskiĭ's theorem 3.33. This construction also allows to describe neighborhoods of points in CAT(0) spaces. This comes from the fact that every CAT(0) space locally looks like a cone over its space of directions. The flat cone construction described in this section is essentially due to [Ber83], see also [BN14].

Similar to flat cones one can define κ-cones, which yield CAT(κ) spaces under suitable conditions. In particular the κ-analog of Theorem 3.33 holds true with some necessary adaptions for $\kappa \neq 0$. We do not cover this generalization here but refer the reader to Section I.5 of [BH99].

Definition 3.28 (The Flat Cone Over a Metric Space) The *flat cone* $C_0(Y)$ over a metric space (Y, d) is, as a set, the quotient of $[0, \infty) \times Y$ by the following equivalence relation:

$$(t, y) \sim (t', y') \text{ if } t = t' = 0 \text{ or if } t = t' > 0 \text{ and } y = y'.$$

We denote the equivalence class of (t, y) by ty and write 0 for the class of $(0, y)$. The latter is also called the *tip* or *vertex* of the cone.

The flat cone should also be a metric space. We hence define the following distance function on it and prove in Lemma 3.31 that it is in fact a metric.

Definition 3.29 (Distance Function on the Flat Cone) Let $C_0(Y)$ be the flat cone over a metric space (Y, d). Put $d_\pi(y, y') = \min\{\pi, d(y, y')\}$ and define a distance function d on $C_0(Y)$ by putting

$$d(ty, t'y')^2 = t^2 + (t')^2 - 2tt' \cos(d_\pi(y, y')).$$

The distance function is constructed so that the distance between ty and 0 is t. Observe that $d(ty, t'y') = t + t'$ if and only if $d(y, y') \geq \pi$.

Example 3.30 The flat cone $X_n := C_0 \mathbb{S}^{n-1}$ is isometric to the Euclidean space of dimension n. To see this, identify $\mathbb{S}^{n-1}$ with the sphere S of radius one around the origin in $\mathbb{R}^n$ via an isometry

$$\sigma : \mathbb{S}^{n-1} \to S.$$

Consider $f : X_n \to \mathbb{R}^n : ty \mapsto t\sigma(y)$. The cosine formula for angles in Euclidean triangles combined with the definition of the distance function on the cone implies that f is an isometry.

The flat cone together with the distance function is indeed always a metric space.

Lemma 3.31 *The distance function d on $C_0 Y$ is a metric.*

Proof It is clear that d is non-negative and symmetric. To see that d satisfies the triangle inequality choose three points $x_i = y_i t_i$, for $i = 1, 2, 3$, in the cone. Consider two cases.
Case (a) One of the t_i equals 0.
Suppose, without loss of generality, that $t_1 = 0$. Then

$$d(x_1, x_2) = t_2 \text{ and } d(x_1, x_3) = t_3.$$

Moreover, since $\cos(a) \leq 1$ for any a, one has

$$d(x_1, x_3) + d(x_3, x_2) = t_3 + \sqrt{t_2^2 + t_3^2 - 2t_2 t_3 \cos(d_\pi(y_2, y_3))}$$

$$\geq t_3 + \sqrt{(t_2 - t_3)^2}$$

$$= t_2 = d(x_1, x_2)$$

and

$$d(x_2, x_1) + d(x_1, x_3) = t_2 + t_3$$

$$\geq \sqrt{t_2^2 + t_3^2 - 2t_2 t_3 \cos(d_\pi(y_2, y_3))}$$

$$= d(x_2, x_3).$$

Case (b) Suppose now that $t_i > 0$ for all i.

Then, as a first subcase, we suppose that $d(y_1, y_2) + d(y_2, y_3) < \pi$. Consider points $\bar{y}_i$, $i = 1, 2, 3$ in $\mathbb{S}^2$ with the property that $d(y_i, y_j) = d(\bar{y}_i, \bar{y}_j)$ for all $i, j \in \{1, 2, 3\}$. One can then show that the cone $C_0\{y_1, y_2, y_3\}$ is isometric to the sub-cone $C_0\{\bar{y}_1, \bar{y}_2, \bar{y}_3\} \subset C_0\mathbb{S}^2 = \mathbb{R}^3$. The triangle inequality of $\mathbb{R}^3$ hence solves this case.

As a second subcase we assume that $d(y_1, y_2) + d(y_2, y_3) \geq \pi$. Consider three points $\bar{y}_i$, $i = 1, 2, 3$, on $\mathbb{S}^1$ such that $d(\bar{y}_1, \bar{y}_2) = d_\pi(y_1, y_2)$ and $d(\bar{y}_2, \bar{y}_3) = d_\pi(y_2, y_3)$ and identify $C_0\mathbb{S}^1$ with $\mathbb{R}^2$ and put $\bar{x}_i = t_i \bar{y}_i$ for $i = 1, 2, 3$. One then concludes that

$$d(x_1, x_2) = d(\bar{x}_1, \bar{x}_2),$$

$$d(x_2, x_3) = d(\bar{x}_2, \bar{x}_3) \text{ and}$$

$$d(x_1, x_3) \leq t_1 + t_3.$$

The last inequality may be seen as follows. Observe that we may replace $-t_1 t_2$ under the square root by the larger value $t_1 t_2$ and combine this with the fact that the cosine function is bounded above by 1. Since we are in the subcase where $d(y_1, y_2) + d(y_2, y_3) \geq \pi$ we may apply Alexandrov's lemma from Exercise 2.41 (see also [BH99, I.2.16]) and obtain that

$$t_1 + t_3 \leq d(\bar{x}_1, x_2) + d(\bar{x}_2, \bar{x}_3).$$

Hence $d(x_1, x_3) \leq d(x_1, x_2) + d(x_2, x_3)$ and the triangle inequality follows. □

We recall, from Theorem I.5.10 in [BH99], an explicit description of the geodesics in flat cones.

Theorem 3.32 (The Flat Cone Theorem) *Let* $x_i = t_i y_i$ *for* $i = 1, 2$ *be two points in* $X = C_0 Y$. *Then the following are true.*

1. *If* $t_1, t_2 > 0$ *and* $d(y_1, y_2) < \pi$, *then there is a bijection between the set of geodesics connecting* y_1 *and* y_2 *in* Y *and the set of geodesic segments connecting* x_1 *and* x_2 *in* X.
2. *In all other cases there is a unique segment joining* x_1 *and* x_2.
3. *Every geodesic segment between* x_1 *and* x_2 *is contained in the closed ball* $\bar{B}_{\max\{t_1, t_2\}}(0)$ *around the cone point* 0 *in* $C_0(Y)$.

The next theorem links the flat cones with the CAT(0) property.

Theorem 3.33 (Berestovskiĭ's Theorem) *The flat cone* $C_0 Y$ *over a metric space* Y *is* CAT(0) *if and only if* Y *is* CAT(1).

Proof Suppose first that $X := C_0 Y$ is CAT(0). Then geodesics in X are unique. The description of geodesics in flat cones in Theorem 3.32 implies that for all $y_1, y_2 \in Y$ with $d(y_1, y_2) < \pi$, there is a unique geodesic connecting them. But then, geodesic triangles are uniquely determined by their vertices.

It thus remains to prove that geodesic triangles with circumference $< 2\pi$ are thinner than their comparison triangles in $\mathbb{S}^2$.

Let $\Delta = \Delta(y_1, y_2, y_3)$ be a geodesic triangle in Y with circumference $< 2\pi$ and let $\bar{\Delta} = \Delta(\bar{y}_1, \bar{y}_2, \bar{y}_3)$ be its comparison triangle in $\mathbb{S}^2$. Pick a point $y \in \operatorname{seg}(y_1, y_2)$ and denote by $\bar{y}$ its comparison point in $\bar{\Delta}$.

Choose $\varepsilon > 0$ and consider the triangle $\Delta_\varepsilon = \Delta(x_1, x_2, x_3)$ where $x_i = \varepsilon y_i$ in the flat cone X. According to Example 3.30 the cone $C_0 \bar{\Delta}$ is a subcone of $C_0 \mathbb{S}^2 \cong \mathbb{R}^3$ and the points $\bar{x}_i = \varepsilon \bar{y}_i$ form a Euclidean comparison triangle, called $\bar{\Delta}_\varepsilon$, for the triangle Δ_ε. The comparison point on $\bar{\Delta}_\varepsilon$ for a point $x = ty \in \operatorname{seg}(x_2, x_3)$ is $\bar{x} = t\bar{y}$. Since we had assumed that X is CAT(0) we may conclude that $d(x_1, x) \le d(\bar{x}_1, \bar{x})$. Therefore, by the definition of the metric on $C_0 Y$, one has that $d(y_1, y) \le d(\bar{y}_1, \bar{y})$ and hence that Δ is thinner than its comparison triangle $\bar{\Delta}$. Thus, Y is CAT(1).

To show the converse suppose that Y is CAT(1). Theorem 3.32 implies that $X = C_0 Y$ is uniquely geodesic.

Consider now three points $x_i = t_i y_i$, $i = 1, 2, 3$, in X.

In case one of the $t_i = 0$, say $t_2 = 0$, the triangle $\Delta(x_1, x_2, x_3) = \Delta(x_1, 0, x_3)$ is isometric to its comparison triangle in $\mathbb{R}^2$. Hence, the triangle is thin in this case and satisfies the CAT(0) condition.

In case that $t_i > 0$ for all i we consider three subcases:

(a) $d(y_1, y_2) + d(y_2, y_3) + d(y_3, y_1) < 2\pi$,
(b) $d(y_1, y_2) + d(y_2, y_3) + d(y_3, y_1) \ge 2\pi$ but $d(y_i, y_j) < \pi$ for all i, j,
(c) $d(y_i, y_j) \ge \pi$ for some i, j.

Case (a) Let $\Delta = \Delta(\operatorname{seg}(y_1, y_2), \operatorname{seg}(y_2, y_3), \operatorname{seg}(y_3, y_1))$ be a triangle in Y and let $\bar{\Delta}$ be its comparison triangle of Δ in $\mathbb{S}^2$ with vertices $\bar{y}_i$. The map from $\bar{\Delta} \longrightarrow \Delta$ defined by $\bar{y} \longmapsto y$ extends to a bijection between $C_0 \bar{\Delta}$ (a subset of $C_0 \mathbb{S}^2 = \mathbb{R}^3$) and $C_0 \Delta$ (a subset of X). One can see that the triangle $\Delta(\bar{x}_1, \bar{x}_2, \bar{x}_3)$ with $x_i := t_i \bar{y}_i$ is the comparison triangle of $\Delta(x_1, x_2, x_3)$.

For an arbitrary point $x = ty$ on $\operatorname{seg}(x_2, x_3)$ and the comparison point $\bar{y}$ of y in $\bar{\Delta}$ one has

$$d(x_1, x) = \sqrt{t_1^2 + t^2 - 2t_1 t \cos(d_\pi(y_1, y))}$$

$$\le \sqrt{t_1^2 + t^2 - 2t_1 t \cos(d_\pi(\bar{y}_1, \bar{y}))}$$

$$= d(\bar{x}_1, \bar{x}).$$

Hence, the CAT(0) condition follows for $\Delta(x_1, x_2, x_3)$ by Alexandrov's result Proposition 3.27.

Case (b) Choose two comparison triangles $\Delta(\tilde{0}, \tilde{x}_1, \tilde{x}_2)$ and $\Delta(\tilde{0}, \tilde{x}_1, \tilde{x}_3)$ of $\Delta(0, x_1, x_2)$ and $\Delta(0, x_1, x_3)$ such that $\tilde{x}_2$ and $\tilde{x}_3$ are on different sides of the edge between the vertices $\tilde{0}$ and $\tilde{x}_1$.

By definition of the cone metric we know that

$$\angle_{\tilde{0}}(\tilde{x}_2, \tilde{x}_3) = 2\pi - \angle_{\tilde{0}}(\tilde{x}_2, \tilde{x}_1) - \angle_{\tilde{0}}(\tilde{x}_3, \tilde{x}_1)$$
$$= 2\pi - (d(y_1, y_2) + d(y_1, y_3))$$
$$\le d(y_2, y_3) \le \angle_0(x_2, x_3)$$

Put $\alpha := \angle_0(x_2, x_3)$ and $\tilde{\alpha} := \angle_{\tilde{0}}(\tilde{x}_2, \tilde{x}_3)$. Note that $\tilde{\alpha}$ is not the angle in the comparison triangle for the triangle with vertices $0, x_2, x_3$ but the angle in the corner $\tilde{0}$ of the two attached smaller triangles.

In a comparison triangle $\bar{\Delta} = \Delta(\bar{x}_1, \bar{x}_2, \bar{x}_3))$ in $\mathbb{R}^2$ of $\Delta(x_1, x_2, x_3)$ one therefore has

$$\angle_{\bar{x}_1}(\bar{x}_2, \bar{x}_3) \ge \angle_{\bar{x}_1}(\tilde{x}_2, \tilde{x}_3)$$
$$= \angle_{\bar{x}_1}(\tilde{x}_2, \tilde{0}) + \angle_{\bar{x}_1}(\tilde{0}, \tilde{x}_3)$$
$$= \angle_{x_1}(0, x_2) + \angle_{x_1}(0, x_3)$$
$$\ge \angle_{x_1}(x_2, x_3).$$

By Alexandrov's characterization of CAT(0) spaces, Proposition 3.27 item 3, the assertion follows.

Case (c) Suppose without loss of generality that $d(y_1, y_3) \ge \pi$. We need to prove that the triangle $\Delta(x_1, x_2, x_3)$ satisfies the CAT(0) condition. Observe that 0 is a point on the geodesic between x_1 and x_3. Construct comparison triangles $\bar{\Delta}_1 = \Delta(\bar{0}, \bar{x}_1, \bar{x}_2)$ and $\bar{\Delta}_2 = \Delta(\bar{0}, \bar{x}_2, \bar{x}_3)$ of the triangle $\Delta_1 = \Delta(0, x_1, x_2)$ and $\Delta_2 = \Delta(0, x_2, x_3)$, respectively. Choose $\bar{\Delta}_1$ and $\bar{\Delta}_2$ in such a way that they share the vertices $\bar{0}, \bar{x}_2$ and such that the points $\bar{x}_1$ and $\bar{x}_2$ are on two different sides of the common geodesic between $\bar{0}$ and $\bar{x}_2$.

Let then α_1 be the angle at $\bar{0}$ between $\bar{x}_1$ and $\bar{x}_2$ and define then α_3 to be the angle at $\bar{0}$ between $\bar{x}_2$ and $\bar{x}_3$. Then

$$\alpha_1 + \alpha_3 = \angle_{\bar{0}}(\bar{x}_2, \bar{x}_1) + \angle_{\bar{0}}(\bar{x}_2, \bar{x}_3) = d_\pi(y_1, y_2) + d_\pi(y_2, y_3) \ge \pi.$$

Here the last inequality follows from the fact that

$$d_\pi(y_1, y_2) + d_\pi(y_2, y_3) \ge d_\pi(y_1, y_3)$$

by the triangle inequality and that $d_\pi(y_1, y_3) = \pi$.

Choose now a comparison triangle $\bar{\Delta}$ for $\Delta(x_1, x_2, x_3)$ and conclude from Alexandrov's Lemma, stated as Proposition 3.27 that the remaining angles in $\bar{\Delta}_1$ and $\bar{\Delta}_2$ are the same as the ones in $\bar{\Delta}$. But then again by 3.27 the claim follows.

Combining these three cases we have shown the CAT(0) condition for X. □

Using the flat cone construction we may now introduce tangent spaces in our metric context.

Definition 3.34 (Tangent Space) For a given metric space X the flat cone over the space of directions at a point $p \in X$ is called *tangent space at* p and denoted by $T_p X := C_0 \Sigma_p X$.

In case that X is a Riemannian manifold the tangent space is isomorphic to the tangent cone at p.

Theorem 3.35 (Curvature of Tangent Things) *Let X be a locally* CAT(κ) *metric space. Then*

1. *The completion of $\Sigma_p X$ is* CAT(1) *for all $p \in X$.*
2. *The completion of $T_p X$ is* CAT(0) *for all $p \in X$.*

Here's a sketch of the strategy for a proof of this theorem: By the theorem of Berestovskiĭ 3.33 it is enough to prove that $T_p X$ is CAT(0). One can also show that it is enough to prove the CAT(0) property locally for a neighborhood of the cone point. One then shows that approximate midpoints (see below) exist and that the CAT(0) 4-point condition is satisfied.

We close this section with one more condition, which allows to characterize CAT(0) spaces.

Definition 3.36 (Approximate Midpoints) A metric space X has *approximate midpoints* if for all $x, y \in X$ and $\varepsilon > 0$ there exists $m \in X$ such that

$$\max\{d(x, m), d(y, m)\} \leq \frac{1}{2} d(x, y) + \varepsilon.$$

Definition 3.37 (The 4-Point-Condition) We say that a metric space X satisfies the CAT(0) *4-point condition* if for all 4-tuples of points $x_1, x_2, y_1, y_2 \in X$ there exist for points $\bar{x}_i, \bar{y}_i \in \mathbb{R}^2$, $i = 1, 2$, such that

$$d(x_i, y_j) = d(\bar{x}_i, \bar{y}_j) \text{ for all } i, j \in \{1, 2\},$$

$$d(x_1, x_2) \leq d(\bar{x}_1, \bar{x}_2), \text{ and}$$

$$d(y_1, y_2) \leq d(\bar{y}_1, \bar{y}_2).$$

Using these properties one can characterize CAT(0) spaces.

3.4 Exercises

Exercise 3.38 Prove that every geodesic triangle Δ in a metric space X has a comparison triangle in M_κ^2. The comparison triangle is unique up to congruence if the circumference of Δ is less than $\frac{2\pi}{\sqrt{\kappa}} = 2D_\kappa$.

Exercise 3.39 Verify the following:

1. $CAT(\kappa)$ implies $CAT(\kappa')$ for all $\kappa' \geq \kappa$.
2. $\mathbb{R}^2$ with the Euclidean metric is CAT(0).

More generally the model space M_κ^2 is CAT(κ). This is, however, much harder to prove.

Exercise 3.40 Let T denote the flat torus modeled by the unit square in $\mathbb{R}^2$ with opposite sides identified. Prove that T is locally CAT(0) but not CAT(0).

Exercise 3.41 Prove that the following are equivalent for a metric realization of a finite graph Γ:

1. Γ is CAT(0).
2. Γ is uniquely geodesic.
3. Γ does not contain any circuits, i.e. sequences $v_0, v_1, \ldots, v_n$ such that v_i is connected to v_{i-1} for all $i = 1, 2, \ldots n$ and v_0 is connected to v_n.

Exercise 3.42 Prove that every convex subset of a CAT(0) space is CAT(0) with respect to the restricted metric.

Exercise 3.43 A *tree* is a graph without circuits, that is a graph without edge paths $v_0, v_1 \ldots, v_k$ where $v_k = v_0$. Prove that geometric realizations of trees, with the length of each edge equal to one, are CAT(κ) for all $\kappa \leq 0$.

So we could say trees are CAT($-\infty$). And in fact one can prove that trees are uniquely characterized by this property.

Exercise 3.44 An $\mathbb{R}$-*tree* is a metric space (T, d) such that

1. there exists a unique geodesic segment $\text{seg}(x, y)$ between every pair of points x, y in T, and
2. if $\text{seg}(y, x) \cap \text{seg}(x, z) = \{x\}$, then the union $\text{seg}(y, x) \cup \text{seg}(x, z)$ is a geodesic.

Show that an $\mathbb{R}$-tree is a CAT(0) space.

The next exercise needs a new definition.

Definition 3.45 (Convex Metric) Let (X, d) be a metric space. Then d is a *convex metric*, or simply *convex*, if for every pair of geodesics $c, c' \colon [0, 1] \to X$ one has

$$d(c(t), c'(t)) \leq (1 - t) \cdot d(c(0), c'(0)) + t \cdot d(c(1), c'(1))$$

for all $t \in [0, 1]$.

Exercise 3.46 Let (X, d) be a CAT(0) space. Show that the metric d is convex.

For a hint see the footnote.[1]

---

[1] To verify the definition of convexity consider the case where $c(0) = c'(0)$ first.

Exercise 3.47 Show that the angle, as defined in Definition 3.20 is, in general, not a metric.

For a hint see the footnote.[2]

Exercise 3.48 Prove that $\Sigma_p X$ is well defined in Definition 3.26 by showing that

$$\gamma \sim \eta :\Leftrightarrow \angle_p(\gamma, \eta) = 0$$

induces an equivalence relation on the set of non-trivial geodesics starting at p.

Exercise 3.49 Let (X, d) be a complete CAT(0) space, $C \subset X$ a closed, convex subset and let $\pi_C : X \to C$ be the projection of X onto C. Prove the following assertions:

1. For all $x, y \in X$ one has $d(\pi_C(x), \pi_C(y)) \leq d(x, y)$.
2. C is a deformation retract of X.
3. Conclude from item 2 that X is contractible.

Exercise 3.50 Let X be complete CAT(0) space, G a group of isometries of X and suppose that the fixed point set X^G is not the empty set. Prove that

1. X^G is closed.
2. X^G is convex if X is uniquely geodesic.

Exercise 3.51 Let (X, d_X) be a metric space and $C_0 X$ the flat cone over X with cone metric denoted by d.

1. Prove that $d_\pi(x, y) := \min\{\pi, d_X(x, y)\}$ is a metric on X.
2. Express the metric d_π on X by the metric d on $C_0 X$.

Exercise 3.52 Let $X = C_0 Y$ be the flat cone with metric d over some metric space Y. Prove that the following statements are equivalent:

1. X is geodesic.
2. Every ball around 0 in X is convex.
3. There exists an open, convex ball around 0 in X.
4. Y is π-geodesic.

Here a metric space (Y, d) is π-*geodesic* if for all $x, y \in Y$ with $d(x, y) < \pi$ there exists a geodesic connecting x and y.

Exercise 3.53 Let $X = C_0 Y$ be the flat cone with cone metric d over the metric space Y. Let $x_1 = t_1 y_1$ and $x_2 = t_2 y_2$ be elements in X with $d(x_1, x_2) = t_1 + t_2$. Show that c defined as follows is a well defined geodesic path from x_1 to x_2:

$$c : [0, t_1 + t_2] \to X, \quad t \mapsto \begin{cases} (t_1 - t)y_1 & 0 \leq t \leq t_1 \\ (t - t_1)y_2 & t_1 \leq t \leq t_1 + t_2 \end{cases}$$

---

[2] Construct an example that does not satisfy positivity.

Chapter 4
Cube Complexes and Gromov's Link Condition

In this chapter, we introduce CAT(0) cube complexes and prove Gromov's characterization of the CAT(0) property within this class of spaces. We start the discussion by introducing a larger class of spaces, called polyhedral cell complexes, in the first section. We then prove the link condition, which reduces the CAT(0) property to checking a curvature condition on the links of vertices of the complexes. The second section contains the definition of a cube complex and the proof of the fact that such a complex is CAT(0) whenever it is simply connected and has flag, simplicial vertex links. Completions of cube complexes are discussed in the third section. The main result here shows that, in essence, non-positive curvature of a cube complex is decided on the 2-skeleton. Finally, in the last section, examples of cube complexes associated with right-angled Artin groups are constructed.

4.1 Polyhedral Complexes and the Link Condition

The class of polyhedral complexes forms a class of spaces generalizing the notions of (metric realizations of) simplicial and cubical complexes at the same time. The building blocks of polyhedral complexes are convex polyhedra, e.g. simplices or cubes, taken from n-dimensional model spaces. The underlying idea is to construct complete geodesic metric spaces by gluing together families of convex polyhedra of varying dimension.

We restrict to spherical and Euclidean convex polyhedra inside the two classes of model spaces defined in 4.1 below. Compare this definition with the 2-dimensional version introduced in Definition 3.1.

Definition 4.1 (Some Model Spaces of Higher Dimension) The *model space* M_0^n, for $n \in \mathbb{N}$, is the Euclidean n-space $\mathbb{E}^n$, that is the set $\mathbb{R}^n$ equipped with the usual Euclidean metric.

© The Author(s), under exclusive license to Springer Nature Switzerland AG 2023
P. Schwer, *CAT(0) Cube Complexes*, Lecture Notes in Mathematics 2324,
https://doi.org/10.1007/978-3-031-43622-2_4

The *model space* M_1^n, again for arbitrary n, is the n-dimensional unit sphere $\mathbb{S}^n \subset \mathbb{R}^{n+1}$ equipped with the angle metric. In this metric the distance between any two points x, y on $\mathbb{S}^n$ is defined to be the angle between the lines through the origin and the points x, respectively, y.

Note that one can also define an induced length metric on $\mathbb{S}^n$ by the Euclidean metric of the ambient $\mathbb{R}^3$. However, this metric is not the right one to be considered in our setting. For further details and a more general treatment of model spaces of arbitrary dimension see Example I.1.2 or Section I.6 in [BH99].

Recall that the *convex hull* of subset A of a metric space (X, d) is the smallest set $\mathrm{conv}(A)$ containing, for all $a, b \in A$, all geodesic segments between a and b.

Definition 4.2 (m-Planes and Hyperplanes in Model Spaces) Consider the model spaces M_κ^n with $\kappa \in \{0, 1\}$. Let $m < n$ be a non-negative integer. An *m-plane* in M_κ^n is the image of an isometric embedding of M_κ^m in M_κ^n. The $(n-1)$-planes are also referred to as *hyperplanes*.

An *m-plane* in $M_0^n \cong \mathbb{R}^n$ is an m-dimensional affine subspace, that is translates of linear subspaces. The m-planes in $\mathbb{S}^n$ are intersections of $\mathbb{S}^n$ with m-dimensional sub-vector-spaces of $\mathbb{R}^{n+1}$.

One can prove, see [BH99] Corollary I.2.22, that every m-plane is the intersection of $(n - m)$ hyperplanes and that every subset S of M_κ^n is contained in a unique m-plane of smallest dimension. This unique m-plane is the intersection of all hyperplanes containing S.

Definition 4.3 (Polyhedral Cells) Fix $\kappa \in \mathbb{R}$. A (convex) M_κ-*polyhedral cell* $C \subset M_\kappa^n$ is a convex hull of a finite set P of points in M_κ^n. In case $\kappa > 0$ we ask P to be contained in an open ball of radius $\frac{1}{2}D_\kappa$. The *dimension* $\dim(C)$ of the cell C is n if C is not contained in any hyperplane of M_κ^n, otherwise $\dim(C)$ is the smallest m such that C is contained in an m-plane. If $\dim(C) < n$ the *interior* C° of C is the interior with respect to a (smallest) m-plane containing C.

Definition 4.4 (Faces, Dimension and Support) Let H be a hyperplane in M_κ^n and let a polyhedral cell C be contained in one of the two closed half-spaces determined by H. If $C \cap H \neq \emptyset$ we call $F := C \cap H$ a *face* of C. If $F \neq C$ we say that F is a *proper* face of C.

The *dimension* k of a face F is the smallest integer such that there exists a k-plane containing F. The 0-dimensional faces of C are called *vertices* of the cell. The *support* $\mathrm{supp}_C(x)$ of a point $x \in C$ is the smallest face of C containing x. By convention the empty set is a face of every cell.

Example 4.5 Let us have a quick look at some (non-)cells.

1. Independent of κ one-cells will be isometric to intervals in $\mathbb{R}$. Moreover, two one-cells constructed from distinct model spaces can be isometric.
2. In dimension two a triangular cell C isometric to an equilateral triangle in $\mathbb{R}^2$ is different from a triangular cell of side lengths 1, 1 and $\sqrt{2}$. While the first has

only one isometry class of faces the latter has two. There are obviously infinitely many isometry classes of 2-cells.
3. Let $\kappa = 1$ and $n = 2$, that is, we are looking at the round unit two-sphere. Consider the set P containing the north pole and two opposite points on the equator. Then P does not span an M_1^2-simplex as P is not contained in an open ball of radius $D_{\frac{\kappa}{2}}$. In particular a closed half-sphere determined by a great circle is not a cell.
4. The support of a vertex of a cell is that vertex. In all other cases the support of a point x in a cell C is strictly larger and x lies in the interior of the face $\text{supp}_C(x)$.

We prove some basic facts about polyhedral cells.

Theorem 4.6 (Properties of Cells) *Let C be an M_κ-polyhedral cell. Then the following are true:*

1. *Every face of C is an M_κ-polyhedral cell of dimension $k \leq n$.*
2. *The intersection of any pair of faces is a face.*
3. *C has finitely many faces.*
4. *C is the convex hull of its vertices.*
5. *If P is a set such that $C = \text{conv}(P)$ then there exists a minimal subset $P' \subset P$ such that $C = \text{conv}(P')$ and P' is the set of vertices of C.*
6. *In case $f : C \to C'$ is an isometry of M_κ-polyhedral cells and F a face of C, then $f(F)$ is a face of C'.*

Proof Let $C = \text{conv}(P)$ be given with P finite. To prove Item 1 let F be a face of C and H a hyperplane such that $F = H \cap C$. The open subspaces determined by H are convex, therefore $F = \text{conv}(P \cap H)$.

Suppose X_i are half-spaces bounded by hyperplanes H_i. The intersection $X_1 \cap X_2$ is then contained in a half-space X_3 bounded by some hyperplane H_3, which satisfies $H_3 \cap (X_1 \cap X_2) \subset (H_1 \cap H_2)$. Hence, Item 2 follows.

Item 3 follows from Item 1. To see this observe that $F = \text{conv}(H \cap P)$ and $H \cap P$ is a proper subset of P.

We leave the proofs of Items 4 and 5 as an exercise.

To prove Item 6 observe first that every isometry between subsets of M_κ^n is a restriction of an element of $\text{Iso}(M_\kappa^n)$. But, every global isometry automatically maps $C \cap H$ for some hyperplane H isometrically onto a different cell C' intersected with some other hyperplane H'. $\square$

Now cells are glued to produce (possibly infinite) complexes.

Definition 4.7 (M_κ-Polyhedral Complex) Let $(C_i, i \in I)$ be a (possibly infinite) family of convex M_κ-polyhedral cells. And let $\sim$ be an equivalence relation on $\mathcal{C} = \bigsqcup_{i \in I} C_i$ that identifies some of the isometric sides of these cells. Consider

$$K = (\bigsqcup_{i \in I} C_i)/\sim$$

with projection map $p : C \to K$ and define maps $p_i : C_i \to K$ by restricting
the projection p to a single cell C_i. Then K is an M_κ-*polyhedral complex* if the
following two conditions are satisfied.

1. For all $i \in I$ and all faces F of C_i the restriction $p_i|_F$ is injective. That is, cells
   are not glued onto themselves.
2. For all $i, j \in I$ and $x \in C_i$, $y \in C_j$ one has: if $p_i(x) = p_j(y)$ then there exists an
   isometry $h : \mathrm{supp}(x) \to \mathrm{supp}(y)$ such that $p_i(z) = p_j(h(z))$ for all $z \in \mathrm{supp}(x)$.

An *n-cell* in K is a subset C of K that is the image of a face F of C_i under p_i. We call
the set of isometry classes of cells the *shapes* of K and denote it with $\mathrm{Shapes}(K)$.
The *dimension* of K is the largest integer n such that there is a shape of dimension n.

For reasons explained in Remark 4.8 we always assume that $\mathrm{Shapes}(K)$ is finite.
This implies that the dimension of the polyhedral complex is finite.

To provide some first examples we note that metric graphs and geometric
realizations of simplicial complexes with Euclidean simplices are M_0-polyhedral
complexes.

Remark 4.8 Allowing for infinitely many different shapes of cells in the definition
of a polyhedral complex can possibly lead to following problem: the intrinsic
pseudo-metric on the complex might not induce a length metric.

Moreover we would like to have that for every point p in the complex there exists
an ε such that the ε-ball around p is (locally) isometric to a κ-cone over the space of
directions. This might again not be the case if we allow for infinitely many isometry
classes of shapes.

We need a couple of more notations and definitions.

Definition 4.9 (Subcomplexes of M_κ-Polyhedral Complexes) Let $(C_i, i \in I)$ be
a family of convex M_κ-polyhedral cells and let $K = (\bigsqcup_{i \in I} C_i)/\sim$ be a polyhedral
complex on this set of cells. A subset K' of K is an M_κ-polyhedral *subcomplex* of
K if there exists a subset $I' \subset I$ such that the subset is itself an M_κ-polyhedral
complex on I', that is

$$K' = (\bigsqcup_{i \in I'} C_i)/\sim$$

with projection map $p' : \bigsqcup_{i \in I'} C_i \to K'$ induced by the projection map of K.

A metric polyhedral complex can be equipped with an intrinsic metric defined
by shortest paths connecting points.

Definition 4.10 (Piecewise Geodesics and the Intrinsic Pseudometric) Suppose
K is an M_κ-polyhedral complex. A given path $\gamma : [a, b] \to K$ is *piecewise geodesic*
if there is a subdivision $x_0 = a < x_1 < \ldots < x_k = b$ of the interval $[a, b]$ and
geodesic paths $c_i : [x_{i-1}, x_i] \to C_{j_i}$ such that $\gamma(t) = p_{j_i}(c_i(t))$ for all i and t. The
length of γ is given by $\ell(\gamma) = \sum_{i=1}^k \ell(c_i)$.

The *intrinsic pseudo-metric d* on K is given by the following formula:

$$d(x, y) = \inf\{\ell(\gamma) \mid \gamma : x \rightsquigarrow y \text{ a piecewise geodesic path in } K\}.$$

One can show that the definition of the length of a path γ is independent of the chosen subdivision of $[a, b]$.

For technical reasons in later proofs we need a constant that encodes for a given point x in a cell C the distance to the closest face F.

Definition 4.11 (Inner Constants) Let K be an M_κ-polyhedral complex. For every point $x \in K$ and every cell C containing x we define a constant $\epsilon(x, C)$ as follows: if $C = \{x\}$ put $\epsilon(x, C) = \infty$ and otherwise let

$$\epsilon(x, C) := \inf\{d_C(x, F) \mid F \text{ a face of } C \text{ and } x \notin F\}.$$

Moreover, let

$$\epsilon(x) := \inf\{\epsilon(x, C) \mid C \text{ a cell with } x \in C\}.$$

We need the next lemma to show that the intrinsic pseudo metric defined in Definition 4.10 is a metric in well behaved complexes. The same lemma is also used in the proof of Theorem 4.19. We leave its proof as an exercise to the reader.

Lemma 4.12 *Let K be an M_κ-polyhedral complex. For every cell C in K denote by d_C the restriction of the metric on the model space to C. Suppose x and y are point in K such that $d(x, y) < \epsilon(x)$. Then every cell C containing y also contains x and $d(x, y) = d_C(x, y)$.*

Corollary 4.13 *In case $\epsilon(x) > 0$ for all $x \in K$ the pseudo-metric defined in Definition 4.11 is a metric and (K, d) is a length space.*

Proof By Lemma 4.12 one has that $d(x, y) > 0$ if and only if $x \neq y$, as this is the case inside C. Therefore d is a length metric by construction. □

Note that this metric coincides with the quotient metric induced by the length metrics on the links of a point x in all cells containing x. Compare [BH99, I.1.7].

4.1.1 The Local Structure of Cell Complexes

We now examine the local structure of polyhedral complexes. The geometry of small neighborhoods has a big impact on the curvature of the space. The main objects that encode this local geometric behavior are stars and geometric links.

Definition 4.14 (Star of a Point) Let K be an M_κ-polyhedral complex. Given $x \in K$ the *star* $\mathrm{St}_K(x)$ *of x in K* is the union of all cells C in K containing x.

We sometimes omit the index K in $\mathrm{St}_K(x)$ and simply write $\mathrm{St}(x)$ if it is clear from the context where the star is taken.

Using the star one can define a space of local directions at any given point. This is done by considering equivalence classes of segments $\mathrm{seg}(x, y)$ for $y \in \mathrm{St}(x) \setminus \{x\}$.

Definition 4.15 (Directions at x) Let K be an M_κ-polyhedral complex and fix x in K. Two points y, y' in $\mathrm{St}_K(x) \setminus \{x\}$ define the same *direction at x* in K if the geodesic segments from x to y and y' satisfy either $\mathrm{seg}(x, y) \subset \mathrm{seg}(x, y')$ or $\mathrm{seg}(x, y') \subset \mathrm{seg}(x, y)$.

The notion of direction induces an equivalence relation on all segments in K. Every equivalence class is contained in a (not necessarily unique) cell.

Definition 4.16 (Geometric Link) Let K be an M_κ-polyhedral complex and fix a point x in K. The *geometric link of x in K* is the set $\mathrm{lk}(x, K)$ of directions at x in K as defined in Definition 4.15. If x is contained in a cell C then $\mathrm{lk}(x, C) \subset \mathrm{lk}(x, K)$ is the set of directions that have a representative $\mathrm{seg}_C(x, y)$ in C.

We often refer to the geometric link at x in K just as *link of x*. In case F is a face of C the link $\mathrm{lk}(x, F)$ is in a natural way a subspace of $\mathrm{lk}(x, C)$. If x is a vertex, then the polyhedral structure on K induces a simplicial structure on the link of x.

For locally finite M_κ-polyhedral complexes the link of a point x can canonically be identified with the space of directions at x as defined in Definition 3.26.

We will now define a metric on links as follows. The main idea is to use angles as distance for directions in a same simplex and to define the distance of arbitrary points u, v of a geometric link $\mathrm{lk}(x, K)$ using approximations by paths, which are piecewise contained in cells.

Following [BH99, Section I.7] we provide an explicit definition of a (pseudo-) metric on links using m-strings.

Definition 4.17 (Metric on Links) Let K be an M_κ-polyhedral complex. We first define distances of directions in the link of a single cell C in K: The *distance* between two directions $u, u' \in \mathrm{lk}(x, C)$ is the usual angle $\angle_C(u, u')$ between the tangent vectors of segments in the equivalence classes u and u'. Let now u, v be arbitrary points in a geometric link $\mathrm{lk}(x, K)$ of $x \in K$. We define their distance via strings. An *m-string* from u to v is a tuple $(u_0, u_1, \ldots, u_m)$ such that $u_0 = u$ and $u_m = v$ and such that for all i there exists a cell C_i with $u_i, u_{i+1} \in \mathrm{lk}(x, C_i)$. The *length* $\ell(\tilde{\Sigma})$ is given by $\ell(\tilde{\Sigma}) := \sum_{i=0}^{m-1} \angle_{C_i}(u_i, u_{i+1})$. And a *pseudo-distance* between u, v is defined via

$$d(u, v) := \inf\{\ell(\tilde{\Sigma}) \mid \tilde{\Sigma} \text{ an } m\text{-string from } u \text{ to } v\}.$$

If there is no m-string connecting u and v in K we put $d(u, v) = \infty$.

Note that the notion of an angle considered in Sect. 3.2 agrees with the usual angle in Euclidean n-space for segments in Euclidean cells.

Fig. 4.1 Measuring distances
of points in a cell of a link

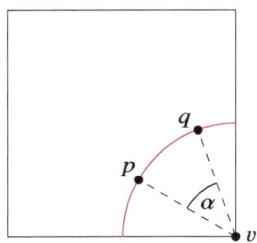

Remark 4.18 The same arguments as in Corollary 4.13 and Lemma 4.12 apply to links with the property that $\epsilon(u) > 0$ for all directions u. Hence, in those cases the pseudo-metric defined in Definition 4.17 is a metric and links are length spaces.

The definition of the metric on the link provided in Definition 4.17 is intuitively best described by measuring the angle between two points in a link $\mathrm{lk}(x, K)$ seen from the vertex x. This is illustrated in Fig. 4.1. For more detailed comments on the link and its metric see [BH99, I.7.15] and the discussions following that item.

We now prove that also neighborhoods of points in M_κ-polyhedral complexes have a well behaved structure.

Theorem 4.19 (Neighborhood of Cone Point) *Let K be an M_κ-polyhedral complex. If $\epsilon(x) > 0$ then the open ball $B_{\frac{\epsilon}{2}}(x)$ in K is isometric to the open ball $B_{\frac{\epsilon}{2}}(0)$ of the cone point 0 in the κ-cone $C_\kappa(\mathrm{l\tilde{k}}(x, K))$.*

Proof We can identify the open ball $B_\epsilon(x)$ in K via a bijective map with the ball of radius ϵ in the κ-cone $C_\kappa(\mathrm{lk}(x, K))$. This is possible by Lemma 4.12 as for all y in $B_\epsilon(x)$ there exists a unique geodesic segment from x to y in a cell C. Let in the following d_C denote the pseudo-metric on C and write d for the cone metric. It then remains to prove that for all y, y' in $B_{\frac{\epsilon}{2}}(x)$ one has $d(y, y') = d_C(y, y')$. To see this let $y = tu$ and let $y' = t'u'$. Lemma 4.12 then implies that $t = d(y, x) = d_C(y, x)$. Suppose that $t, t' > 0$, then triangle inequality implies that both distances $d(y, y')$ and $d_C(y, y')$ are less than or equal to $t + t'$. ☐

We first prove the following two claims.

Claim 1 *If $d(u, u') < \pi$ then $d(y, y') \leq d_C(y, y') \leq t + t'$.*

Claim 2 *If $d(y, y') < t + t'$ then $d(u, u') < \pi$ and $d_C(y, y') \leq d(y, y')$.*

Proof of claim 1: By definition the distance between u and u' is the infimum over lengths of chains from u to u'. Hence, there exists a chain $\tilde{\Sigma} = (u = u_0, \ldots, u_m = u')$ such that $d(u, u') \leq \ell(\tilde{\Sigma}) < \pi$. Choose cells C_i for all i such that $u_i, u_{i+1} \in \mathrm{lk}(x, C_i)$. We construct a chain Σ connecting y and y' such that the radial projection from Σ onto $\mathrm{lk}(x, K)$ is $\tilde{\Sigma}$.

Here the radial projection is the map that sends an element $y \in \mathrm{St}(x) \setminus \{x\}$ to the equivalence class of the geodesic from x to y. This map sends open simplices to open simplices.

Fig. 4.2 Shortcut of a chain
which illustrates a step in the
proof of Theorem 4.19

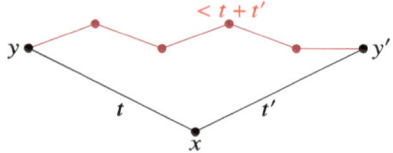

The construction of the chain Σ is now as follows. Fix a basepoint $\bar{x}$ in M_κ^2 and choose points $\bar{x}_0$ and $\bar{x}_m$ in M_κ^2 such that $d(\bar{x}, \bar{x}_0) = t$, $d(\bar{x}, \bar{x}_m) = t'$ and the angle $\angle_{\bar{x}}(\bar{x}_0, \bar{x}_m) = \ell(\tilde{\Sigma})$. Then choose points $\bar{x}_i$ for $i = 1, 2, \ldots, m - 1$ on the geodesic ray from $\bar{x}$ through u_i such that $\angle_{\bar{x}}(\bar{x}_i, \bar{x}_{i+1}) = d_{\mathrm{lk}(x, C_i)}(u_i, u_{i+1})$. Put $t_i = d(\bar{x}, \bar{x}_i)$ and let $x_i = t_i u_i \in B_{\frac{\epsilon}{2}}(x)$. The geodesic segment between x_i and x_{i+1} is then contained in C_i by construction and hence $d_{C_i}(x_i, x_{i+1}) = d(\bar{x}_i, \bar{x}_{i+1})$. We collect all the vertices we have just constructed in a chain $\Sigma = (x_0, x_1, \ldots, x_m)$. This chain connects $y = x_0$ with $y' = x_m$ in $B(x, \varepsilon(x))$. We therefore have the following equation:

$$d(y, y') \leq \ell(\Sigma) = \sum_i d_{C_i}(x_i, x_{i+1}) = d(\bar{x}_0, \bar{x}_m).$$

But then, $d_C(y, y') = \inf_{\tilde{\Sigma}} \ell(\tilde{\Sigma}) = \inf_{\tilde{\Sigma}} d(\bar{x}_0, \bar{x}_m) \geq \ell(\Sigma) \geq d(y, y')$ and Claim 1 follows.

To prove claim 2 we let $\Sigma = (x_0, \ldots, x_m)$ be a chain connecting $y = x_0$ with $y' = x_m$ in $B_\epsilon(x)$. Let $\ell(\Sigma) < t + t'$. The triangle inequality then implies that x does not lie on the path defined by Σ. We may hence write $x_i = t_i u_i$ and define a chain $\tilde{\Sigma} = (u_0, \ldots, u_m)$ in $\mathrm{lk}(x, K)$ connecting $u = u_1$ with $u' = u_m$. Compare Fig. 4.2.

As above we develop $\tilde{\Sigma}$ to a chain $\bar{\Sigma}$ in M_κ^2 and suppose that $\ell(\tilde{\Sigma}) > \pi$. Then there exists an index $0 < i < m$ and a point $\bar{x}_i'$ on the geodesic segment between $\bar{x}_i$ and $\bar{x}_{i+1}$ such that $\angle_x(\bar{x}_0, \bar{x}_i') = \pi$. Put $t_i' := d(\bar{x}, \bar{x}_i')$. See Fig. 4.3 where this is illustrated.

We then put $\bar{\Sigma}' := (\bar{x}_0, \ldots, \bar{x}_i, \bar{x}_i')$ and $\bar{\Sigma}'' := (\bar{x}_i', \bar{x}_{i+1}, \ldots, \bar{x}_m)$ and obtain

$$\ell(\bar{\Sigma}) = \ell(\bar{\Sigma}') + \ell(\bar{\Sigma}'') \geq (t_0 + t_i') + (t_m - t_i') = t + t',$$

which yields a contradiction.

Fig. 4.3 Comparing angles
in a step of the proof of
Theorem 4.19

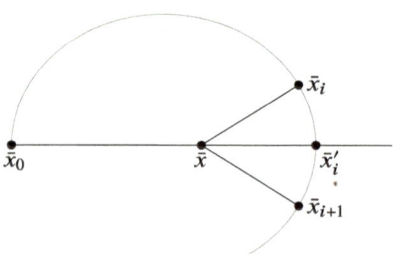

Fig. 4.4 This figure shows (in red) the link of a vertex x' in an M_0-cubical complex. The neighborhood of the point x has a piece of dimension two (inside the square) and a piece of dimension three (inside the 3-cube)

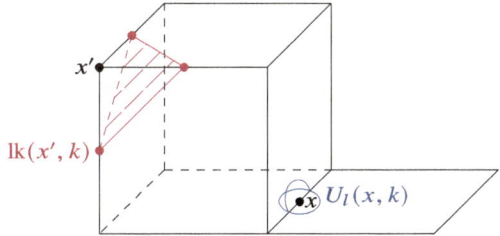

One may hence assume that $\pi > \ell(\tilde{\Sigma})$ and concludes

$$\pi > \ell(\tilde{\Sigma}) = \angle_{\bar{x}}(\bar{x}_0, \bar{x}_m) \geq d(u, u_m) = d(u, u').$$

The distances $d(\bar{x}_0, \bar{x}_m)$ and $d_C(y, y')$ can then be computed via the cosine rule for triangles in M_κ^2 via the angle $\angle_{\bar{x}}(\bar{x}_0, \bar{x}_m)$ and $d_C(u, u')$. One then obtains that $\ell(\Sigma) \geq d(\bar{x}_0, \bar{x}_m) \geq d_C(y, y')$ and hence $d(y, y') = \inf \ell(\Sigma) \geq d_C(y, y')$ and the second claim follows.

To complete the proof of the theorem suppose now that $d(y, y') = t + t'$. Then $d(u, u') \geq \pi$ by Claim 1. Therefore $d_C(y, y') = t + t'$ and the statement of the theorem follows by Claim 2.

Figure 4.4 illustrates links and ϵ-neighborhoods in an M_κ-polyhedral complex with cubical cells.

Gromov established a criterion that allows to test a space for local non-positive curvature by examining the curvature of its vertex links. We prove this result as Theorem 4.22 and introduce the following definition that comes with it next.

Definition 4.20 (The Link Condition) We say that an M_κ-polyhedral complex X satisfies the *link condition* if all its vertex links are CAT(1).

We need one preparatory lemma for the proof of Theorem 4.22. Its proof can be found in [BH99, I.7.56.]

Lemma 4.21 *Let K be an M_κ-polyhedral complex with finitely many shapes. Let x, y be points contained in a same open cell in K. Then, for ϵ small enough, there exists an isometry $f : B_{\frac{\epsilon}{2}}(x) \to B_{\frac{\epsilon}{2}}(y)$ such that for every closed cell C containing x the restriction $f|_C$ is an isometry from $B_{\frac{\epsilon}{2}}(x) \cap C$ to $B_{\frac{\epsilon}{2}}(y) \cap C$.*

We are now ready to prove that the link condition does indeed characterize upper curvature bounds.

Theorem 4.22 (The Link Condition) *Let X be an M_κ-polyhedral complex with finitely many shapes. Then X has curvature $\leq \kappa$ if and only if X satisfies the link condition.*

Proof Let v be a vertex in X. By Theorem 4.19 there exists $\epsilon > 0$ such that $B_{\frac{\epsilon}{2}}(x)$ is isomorphic to a neighborhood of the conepoint in the κ-cone $C_\kappa(\mathrm{lk}(v, X))$. Berestovskiĭ's Theorem 3.33 then implies that X satisfies the link condition if and only if every vertex contains a neighborhood that is CAT(κ). Thus, the assertion follows from Lemma 4.21.   □

Please note that the condition that X in Theorem 4.22 has finitely many shapes can be weakened to the assumption that X is finite dimensional and for all $x \in X$ the constant $\epsilon(x)$ is strictly positive.

4.2 CAT(0) Cube Complexes

Cube complexes are spaces obtained by gluing cubes of various dimensions along isometric faces. In short, a *cube complex* X is an M_0-polyhedral complex in which all the shapes are unit cubes, i.e., of the form $[0, 1]^k$ for some $k \in \mathbb{N}$, and such that no cube is glued to itself and the intersection of two cubes is always a face of both. In this section we provide an ad-hoc definition of cube complexes in order to allow for a treatment of the subject without the need of introducing general M_κ-polyhedral complexes. Afterwards we study some of their elementary properties and prove, as a highlight of this section, the cubical version of Gromov's link condition, that is Theorem 4.43.

We start with the definition of a cube.

Definition 4.23 (Cubes) The set $C_n = [0, 1]^n \subset \mathbb{R}^n$ is called *standard n-cube*, short *n-cube*. We say that C is a *(standard) cube* if C is a standard n-cube for some n. A *codimension one face* of an n-cube $C = C_n$ is a subset $F \subset C$ such that $F = F_{i,\varepsilon}$ for some choice of $\varepsilon \in \{0, 1\}$ and $i = 1, \ldots, n$, where $F_{i,\varepsilon}$ is given by

$$F_{i,\varepsilon} := \{x \in C \mid x_i = \varepsilon\}.$$

All other *(proper) faces* of C are non-empty intersections of codimension one faces. We say that x is an *inner point* of C if x is not contained in any (proper) face of C.

An example of a cube and some of its faces can be found in Fig. 4.5. It is not hard to check that the poset of faces ordered by inclusion is independent of the chosen isometry of C to the standard n-cube.

Next we define cube complexes. For technical reasons we assume that the intersection of two cells in a cube complex is either empty or a face of both. Some, but not all, of the gluings that do not satisfy this assumption can be resolved by further subdividing the complex into smaller cubes as illustrated in Example 4.45.

Definition 4.24 (Cube Complexes) Let C and C' be two standard cubes with faces F and F', respectively. A *gluing* of C and C' is an isometry $\phi : F \to F'$.

Suppose $\mathcal{C}$ is a set of standard cubes and $\mathcal{S}$ a family of gluings of elements of $\mathcal{C}$, that is, for all $C \in \mathcal{C}$ there is $n_C \in \mathbb{N}$ such that $C \cong [0, 1]^{n_C}$ and every $\phi \in \mathcal{S}$ is an isometry $\phi : F \to F'$ where F, F' are faces of cubes $C, C' \in \mathcal{C}$. Let $\sim$ be the equivalence relation on the disjoint union $\bigsqcup_{C \in \mathcal{C}} C$ of all cubes that is generated by

Fig. 4.5 Some faces of a 3-cube

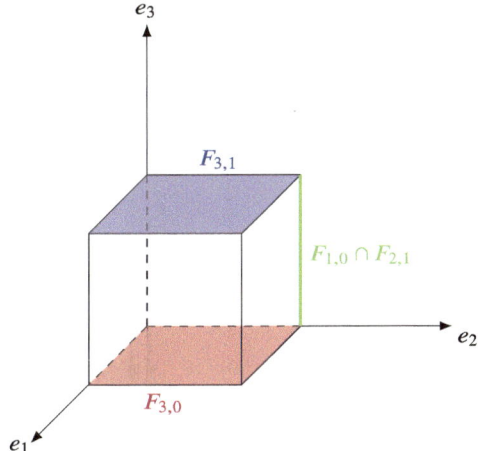

$x \sim \phi(x)$ for $\phi \in \mathcal{S}$ and $x \in \mathrm{dom}(\phi)$. Assume that the following conditions are satisfied

1. For all $C \in \mathcal{C}$ and all $x \in C$ there does not exists $y \in C$ such that $y \neq x$ and $y \sim x$.
2. For all distinct cubes $C, C' \in \mathcal{C}$ there is at most one gluing of C and C'.

Then the pair $(\mathcal{C}, \mathcal{S})$ defines the underlying set X of a metric *cube complex* (X, d) by putting

$$X := \left(\bigsqcup_{C \in \mathcal{C}} C \right) \Big/ \sim \qquad (4.1)$$

We refer to the images of the standard cubes $C \in \mathcal{C}$ in X as *cubes in X* and will also speak of their dimension, faces, etc.. The metric d on X is the length metric induced by the (restricted) Euclidean metric on each cube in $\mathcal{C}$. The *dimension* of X is the largest integer n such that $\mathcal{C}$ contains an n-cube. A *combinatorial map* between two cube complexes is a map $f : X \to Y$ that sends k-cubes of X to k-cubes of Y for all $0 \leq k \leq n := \dim(X)$.

Note that one can glue cubes which are not of a same dimension. However, the faces that are identified by a gluing do have the same dimension. One may, for example, glue a 2-cube onto a 3-cube along some edge.

Condition 1 of Definition 4.24 implies that no pair of points inside a cube are identified by the gluings. Loosely speaking no cube is glued to itself. By condition 2 of the same definition one immediately obtains that any pair of distinct cubes is glued along at most one face. We make these observations precise in the next proposition, which is an easy consequence of Definition 4.24. The assertion in the first item of the proposition allows us to identify a cube $C \in \mathcal{C}$ with its image in X and speak of a *cube C in X*. We sometimes write $C \in X$.

Proposition 4.25 (Properties of a Cube Complex) *For a cube complex X, defined by a pair $(\mathcal{C}, \mathcal{S})$ of a set of cubes and a family of gluing, the following is true.*

1. For every cube $C \in \mathcal{C}$ the restriction of the quotient map

$$\mathrm{p} : \bigsqcup_{C \in \mathcal{C}} C \to X$$

to C yields an injective map $\mathrm{p}|_C : C \to X$.
2. The intersection of two cubes in X is either empty or a face of both (here a face might be the whole cube).

Convince yourself that in a cube complex links, as defined in 4.16, of cubical cells and in particular of vertices are simplicial complexes. Some sources, e.g. [HW08], do not assume that gluings satisfy conditions 1 and 2 in Definition 4.24 and call any quotient by gluings a cube complex. Such a complex is then called *simple* if vertex links are simplicial complexes. Using the definition above every cube complex is simple in the sense that all its (vertex) links are simplicial complexes. This simplifies some of the statements and proof in later sections.

Definition 4.26 (Skeleta of Cube Complexes) The k-*skeleton* $X^{(k)}$ of a cube complex X is the subset of X consisting of all cubes of dimension $\leq k$ in X. It is hence (as a set) the union

$$X^{(k)} = \bigcup_{C \in \mathcal{C}_k} C \text{ with } \mathcal{C}_k = \{C \in \mathcal{C} \mid C = C_i \text{ for some } i \leq k\}.$$

A first example of a cube complex and its 1-skeleton is given below.

Example 4.27 A cube complex consisting of a single 3-cube with two 2-cubes attached to two of its vertical edges is shown on the left-hand side of Fig. 4.6. Its 1-skeleton is shown on the right.

Remark 4.28 In the definition of cube complexes we have assumed that the intersection of any two cubes should either be empty or a cube as well. Examples for which this assumption is not satisfied are the tube, a cornet and the pillowcase. By a *tube* we mean a self-gluing of a cube along opposite faces. A *cornet* is a self-gluing of a 2-cube along two of its edges sharing a vertex. Finally in order to construct a

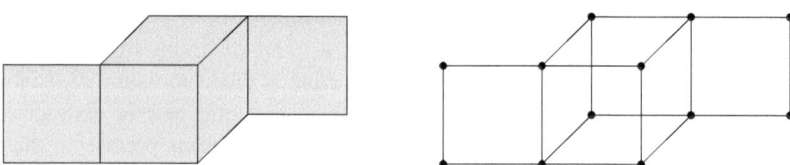

Fig. 4.6 A cube complex on the left with its 1-skeleton on the right

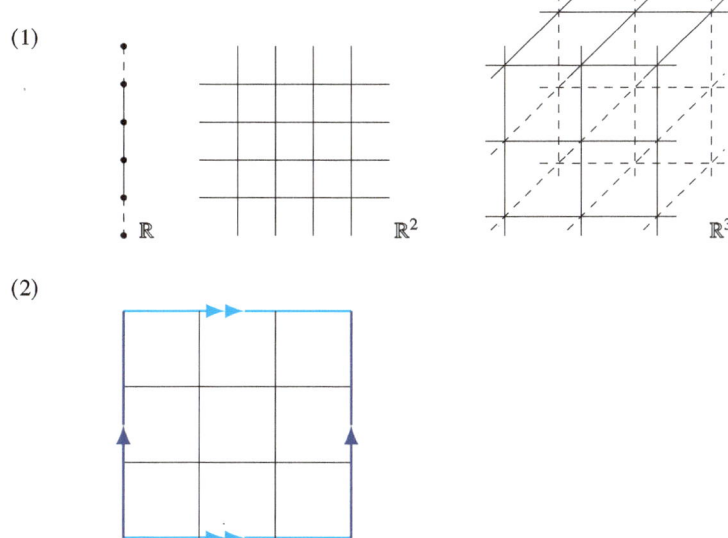

Fig. 4.7 Examples of cube complexes: standard cubing of Euclidean spaces in (1) and a cubulated 2-torus in (2)

pillowcase take two 2-cubes and glue a pair of edges sharing a vertex in one cube to another pair of edges sharing a vertex in the second cube. The right-hand side of Fig. 5.14 shows such a pillowcase.

The cube forming a tube (via the self-gluing explained above) can be barycentrically subdivided into smaller cubes such that the gluing yields a cube complex. This means that the self-gluing of this kind can be resolved by considering a finer cubical structure.

However, neither the problematic situation of the cornet nor that of the pillowcase can be resolved using cubical barycentric subdivision. Suppose we have a cornet or a pillowcase and let v denote the vertex whose link is not a simplicial complex. The cubical barycentric subdivision of a 2-cube adds a vertex v_m in the midpoint of the 2-cube and a vertex in the midpoint of each of its edges. The vertex v_m is then connected by a (new) edge to each vertex at a midpoint of the four original edges. Doing this the link of v will not change.

We continue with some more examples of cube complexes.

Example 4.29

1. Metric realizations of graphs in which all edges have length one are cube complexes. Each edge is a 1-cube of the complex. The codimension one faces of the cubes are the vertices. Each edge (i.e. 1-cube) is glued to its neighboring edges along a vertex. In particular, trees are cube complexes.

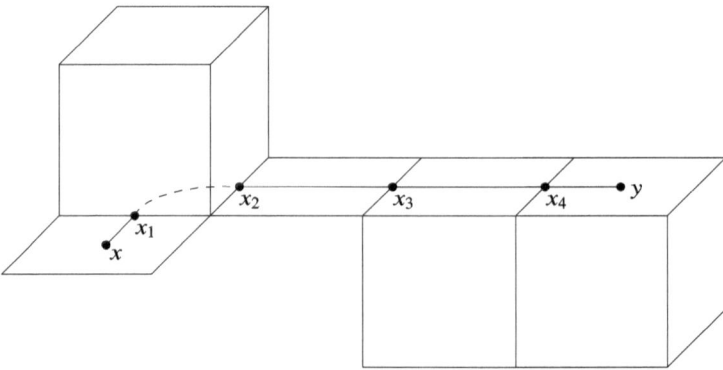

Fig. 4.8 This figure shows a string connecting points x and y in a string-connected cube complex

2. For any n the Euclidean n-space $\mathbb{R}^n$ carries a natural structure of an n-dimensional cube complex. The *standard cubing* of $\mathbb{R}^n$ is such that each subset of the form

$$\{(x_1, x_2, \ldots, x_n) \mid a_i \leq x_i \leq a_i + 1\}, \ a_i \in \mathbb{Z} \text{ for all } i$$

is the image of a cube. Figure 4.7 shows in (1) the standard cubings of $\mathbb{R}^n$ for $n = 1, 2, 3$.
3. The flat 2-dimensional torus carries a cubical structure. Take the subset $\{(x, y) \in \mathbb{R}^2 \mid 0 \leq x, y \leq 3\}$ of $\mathbb{R}^2$ with the (restricted) standard cubing and identify opposite sides of this square. Compare 4.7 picture (2) with lines with same arrows identified.

Cube complexes, as a sub-class of polyhedral complexes, carry a natural polyhedral metric as defined in Definition 4.10. We repeat the definition of the distance function in this special case and prove that it is indeed a metric in case of cube complexes. See Proposition 4.32 below.

Notation 4.30 (Strings and Distance Function) Let x and y be two points in a cube complex X. A *string* Σ from x to y is a sequence of points $x_i, i = 1, \ldots, m$ such that $x_0 = x$, $x_m = y$ and for all $i = 0, \ldots, m$ there exists a cube C_i containing x_i and x_{i+1}.
The *length* of a string Σ is given by $l(\Sigma) := \sum_{i=0}^{m-1} d_{C_i}(x_i, x_{i+1})$ where d_{C_i} is the Euclidean metric on C_i.
Define a *distance function* $d : X \times X \to \mathbb{R}$ by

$$d(x, y) := \inf\{L(\Sigma) \mid \Sigma \text{ is a string from } x \text{ to } y\}.$$

Note that if X is connected there exists for every pair of points x, y in X a string from x to y. The length of a string is well defined by item 2 of Definition 4.24. An example of a string can be found in Fig. 4.8.

We need one more technical lemma for the proof of Proposition 4.32. Recall from Definition 4.11 that the constant $\epsilon(x, C)$ for a point x contained in a cube C is given by

$$\epsilon(x, C) = \inf\{d_C(x, F)| \ F \subset C \text{ and } x \notin F\}.$$

Lemma 4.31 *Let x be a point in a connected cube complex X of dimension at least 1. Let further C be a cube containing x. Then $\epsilon(x, C)$ is independent of the choice of the cube C.*

Proof If for a fixed x there is a unique cube containing x in its interior, then the assertion is clear. In case that x is a vertex of C (that is, a face of dimension 0), then x is a vertex in every cube containing it and $\epsilon(x, C) = 1$.

Suppose now that $x \in C$ be neither an interior point of C nor a vertex. Then let $F_C \subset C$ be the face of minimal dimension such that $x \in F_C$. Such a face exists for every cube containing x and its dimension is always the same. The (unique) point $y \in C$ with $d(x, y) = \epsilon(x, C)$ is contained in a face of F_C. Thus

$$\epsilon(x, C) = \inf\{d_{F_C}(x, G)|G \text{ is a face of } F_C\}.$$

Since F_C and $F_{C'}$ need to be isometric for cubes C, C' containing x the claim follows.   □

Proposition 4.32 (The Distance Function Is a Metric) *For any cube complex X the distance function $d : X \times X \to \mathbb{R}$ defined in Notation 4.30 is a metric. Moreover, for all $x, y \in X$ one has*

$$d(x, y) = \inf\{l(\gamma)| \ \gamma \text{ is a rectifiable curve } x \rightsquigarrow y\}$$

and d is a length metric.

Proof Symmetry of d follows directly from the definition by reading strings backwards. Connecting strings $\Sigma : x \rightsquigarrow y$ and $\Sigma : y \rightsquigarrow z$ in series gives us strings from x to z and hence the triangle inequality holds. Positivity of d comes from the fact that if a cube C contains x then it contains every y with $d(x, y) < \epsilon(x, C)$. Hence, in this case $d(x, y) = d_C(x, y)$ where d_C is the (restricted) Euclidean metric on C. The fact that d is indeed a length metric is a direct consequence of the definition of d.   □

Definition 4.33 (Finite Dimensional, Locally Finite) We say that a cube complex (X, d) defined by the pair $(\mathcal{C}, \mathcal{S})$ is *finite dimensional* if there is a global upper bound on the dimension of cubes in $\mathcal{C}$. It is *locally finite* if no point of X is contained in infinitely many cubes.

Finite dimensionality and locally finiteness guarantee that a cube complex has a well behaved metric structure.

Proposition 4.34 (Complete Geodesic Cube Complex) *A cube complex is a complete geodesic metric space if it is finite dimensional or locally finite.*

We leave the proof of this proposition as an exercise. A reference in case of finite dimensional X is [BH99, Thm. I.7.19]. If X is locally finite show first that X is proper (using the Hopf-Rinow Theorem 2.17) and show then that X is geodesic.

Example 4.35 Here is a first example of an infinite dimensional cube complex: Let $B = \{b_1, b_2, \ldots\}$ be an ordered basis of the infinite dimensional vector space $H = \bigoplus_\mathbb{N} \mathbb{R}$. Take in H the ascending union of the unit cubes spanned by the first k basis elements of B, where k ranges over $\mathbb{N}$. The resulting cube complex contains infinitely many cubes of increasing dimension. In particular, there is no global bound on the dimension of its cubes. Moreover, the complex is not complete.

Remark 4.36 There are a priori two topologies on a cube complex X. For one the quotient topology $\mathcal{T}_p$ and on the other hand the topology $\mathcal{T}_d$ induced by the length metric. One has, in general, $\mathcal{T}_d \subset \mathcal{T}_p$ and equality if and only if X is locally finite.

We now prove a criterion, which allows us to easily characterize the those cube complexes that are CAT(0) via a simple property of their links. This beautiful criterion was established by Gromov and is one of the main reasons why cube complexes are so popular in geometric group theory. In general, curvature testing is hard and even for simplicial complexes no good characterizations are known. So it is remarkable how easy testing for CAT(0) in the class of cube complexes is.

Definition 4.37 (All-Right Spherical Shapes and Complexes) We consider the following shapes S^n, which are right angled cutouts of spheres. That is, for $n \geq 1$ fix a cube $C^{n+1} \cong [0, 1]^{n+1}$ and some $\varepsilon > 0$. Let v be a corner of C, then the *all-right spherical shape* S^n of dimension n is given by

$$S^n := \{y \in C \mid d(v, y) = \varepsilon\}.$$

Distances of points p, q in S^n are measured in the angular metric as follows: $d_{S^n}(p, q) := \angle_v(\text{seg}(v, p), \text{seg}(v, q))$, where $\angle_v$ stands for the Euclidean angle at v in C. The *faces* of S^n are the intersections of S^n with faces of C (and are isomorphic to some S^k with $k < n$.

An *all-right spherical complex* is a polyhedral complex built out of all-right spherical shapes satisfying gluing rules analogous to the ones given in Definition 4.24 with $\mathcal{C}$ replaced by a family $\mathcal{S}$ of all-right spherical shapes.

Observe that links of all-right spherical complexes are again all-right spherical complexes. We sometimes write *spherical complex* when we actually mean an all-right spherical complex, since these are the only ones appearing in this book.

Examples of all-right spherical shapes are shown in Fig. 4.9. See also Fig. 4.10 where all-right spherical complexes appear as links.

Links of vertices in a cube complex can either be defined via Definition 4.16 or using the following equivalent characterization.

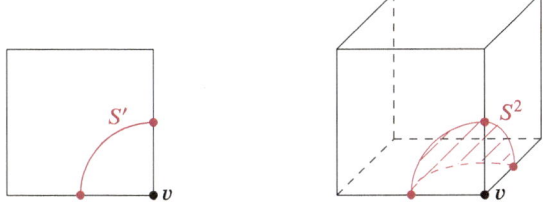

Fig. 4.9 All-right spherical simplices appear as links in cubes

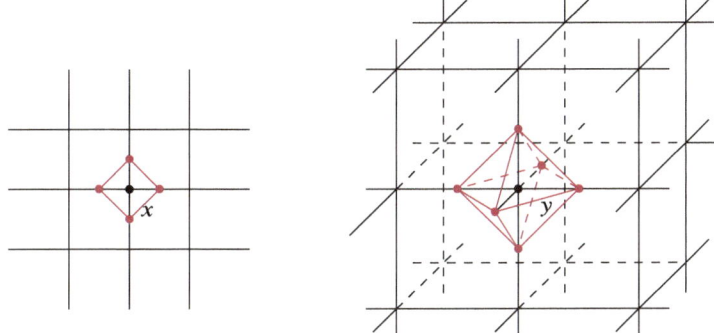

Fig. 4.10 The red graphs illustrate the 1-skeleta of vertex links of x, y in the standard cubings of $\mathbb{R}^2$ (left) and $\mathbb{R}^3$ (right)

Definition 4.38 (Links in Cube Complexes) Let $1 > \varepsilon > 0$. The *link* lk(v, X) of a vertex v in a cube complex X is the spherical complex obtained from the ε-sphere around v in X equipped with the simplicial structure induced from X.

By the assumption that cubes are not glued to themselves and no two cubes share more than one (maximal) face we may conclude that links carry in a natural way the structure of an abstract simplicial complex. This is obtained by forgetting the metric and the spherical nature of its simplices.

Example 4.39 First consider the standard cubings of the Euclidean plane and 3-space. Since these tilings are regular, the links of all their vertices look the same. See Fig. 4.10 for an illustration of the cubings and links. In dimension two all vertex links are 1-spheres consisting of four edges. In dimension three the link of every vertex is an octahedron built from all-right-spherical cells.

Further examples of links in cube complexes can be found in Fig. 4.11. Here links may look different. The link of the vertex labeled x consists of a spherical 1-cell glued onto a spherical 2-cell. The link of the vertex y is a cycle consisting of four edges.

The next property is crucial for the CAT(0) characterization of cube complexes.

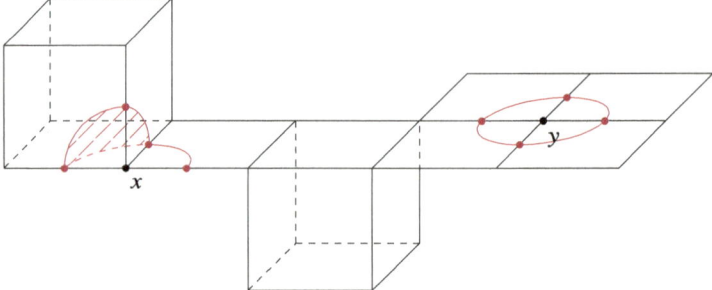

Fig. 4.11 Links in a cube complex are all-right spherical complexes

Definition 4.40 (Flag Simplicial Complex) An abstract simplicial complex Δ is *flag* if and only if every subset of the vertices of Δ that spans a complete graph in the 1-skeleton also spans a simplex.

A simplicial complex is flag if there are no empty simplices. That is, whenever the 1-skeleton of a simplex is seen the simplex itself exists. The links in the standard cubing of the Euclidean plane and 3-space are flag simplicial complexes. See Example 4.39.

We show that flagness translates into a metric curvature condition. The proof uses the following curvature criterion for spherical complexes.

Proposition 4.41 (Testing Curvature) *A spherical complex Δ is CAT(1) if and only if it is locally CAT(1) and there are no locally geodesic circles of length $< 2\pi$.*

This proposition is true in a slightly more general setting. For a precise statement and proof see Theorem I.5.4 in [BH99].

Proposition 4.42 (Flagness Versus the CAT(1) Condition) *A finite dimensional spherical complex is CAT(1) if and only if it is a flag simplicial complex.*

Proof Let D denote the spherical complex. Suppose that D is CAT(1). We prove flagness by induction on the size of a clique, or empty simplex, in the 1-skeleton.

For $n = 1$ there is nothing to prove. So let $n = 2$. We then need to show that there are no empty triangles. Suppose for a contradiction that there is an empty triangle $\Delta(x, y, z)$. In this empty triangle the distance from y to the midpoint m of the side on x and z is $\frac{3}{4}\pi$. The distance between the comparison points $\bar{y}$ and $\bar{m}$ of y and m in the spherical comparison triangle is smaller than $\frac{3}{4}\pi$. But, this contradicts the CAT(1) property of D.

We proceed with the induction step from $(n - 1)$ to n. Suppose the assertion is true for $n - 1$. Let $\{v_0, \ldots, v_n\}$ be vertices in D that span the $(n - 1)$-skeleton S of an n-simplex in D. By the induction hypothesis, as the dimension of the link is $n - 1$, we obtain for all vertices $v = v_i, i = 0, \ldots, n$ that the link $\mathrm{lk}(v, D)$ is CAT(1) and a flag simplicial complex. But then the intersection of the link with S is the $(n - 2)$-skeleton of an $(n - 1)$-simplex in $\mathrm{lk}(v, D)$, which is in fact the boundary of an $(n-1)$-simplex. This implies that S is the boundary of an n-simplex and D is flag.

To show the reverse implication we proceed by induction on the dimension of D. Suppose first $\dim(D) = 1$. The triangles in a one-dimensional flag simplicial complex with circumference less than 2π are automatically degenerate and hence CAT(1).

For the induction step suppose that $\dim(D) = n$ and that D is flag. Suppose further that the assertion holds for $(n-1)$-dimensional complexes. To prove the claim it is enough to show that there are no geodesic loops of length less than 2π. Since we are dealing with a right-angled spherical complex one can conclude that

$$\mathrm{lk}(v, D) \cong S_{\frac{\pi}{2}} = \left\{ x \in D \mid d(x, v) = \frac{\pi}{2} \right\}.$$

Let γ be a locally geodesic loop. Let v be a vertex in D such that the intersection $c := \mathrm{Im}(\gamma) \cap B_{\frac{\pi}{2}}(v)$ is not empty.

We claim that c is a curve of length π. In case that $v \in c$ the claim follows. In case $v \notin c$ we now construct a surface S in $\overline{B_{\frac{\pi}{2}}(v)} = \{x \in D \mid d(x, v) \leq \frac{\pi}{2}\}$. The surface S is defined to be the union of all geodesics of length $\frac{\pi}{2}$ starting in v and containing at least one point of c. One then has that S is the union of spherical triangles D_i of height $\frac{\pi}{2}$ sitting along c. See Fig. 4.12 for an example how this might look like.

We subdivide c by points x_i defined by the faces of these triangles cutting c. The the ball $B_{\pi/2}(v)$ is isomorphic to the CAT(1) cone over $S_{\frac{\pi}{2}}(v)$. This implies that there is an embedding ι from S to $\mathbb{S}^2$ such that the restriction of ι onto each triangle D_i is an embedding with the property that c is mapped to a local geodesic. Assume without loss of generality that $\iota(v)$ is the north pole of the sphere. Then the image $\iota(c)$ connects two points on the equator and intersects the interior of the upper hemisphere non-trivially. Hence, $\ell(\iota(c)) = \pi$ and thus also $\ell(c) = \pi$ as local

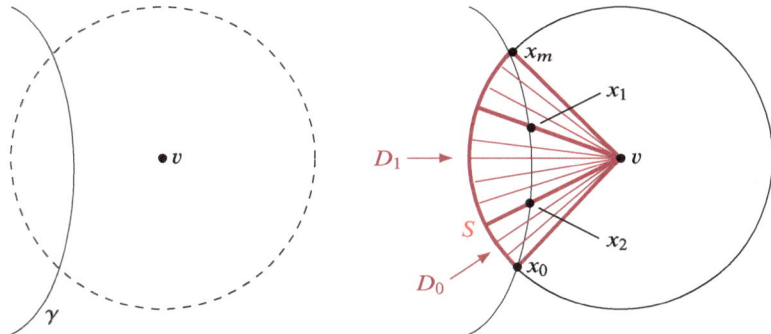

Fig. 4.12 An illustration of the surface spanned by a curve in the proof of Proposition 4.42

geodesics of length up to π in $\mathbb{S}^2$ are also global geodesics.[1] We have thus shown the claim that the length of c is π.

For the last step of the proof we suppose that $\ell(\gamma) < 2\pi$. But then, γ can not be the concatenation of two curves of length equal to π (obviously) and can hence not be contained in the union of two balls $\bar{B}_{\frac{\pi}{2}}(v)$ and $\bar{B}_{\frac{\pi}{2}}(u)$ where $B_{\frac{\pi}{2}}(v) \cap B_{\frac{\pi}{2}}(u) = \emptyset$.

We hence conclude that the set $B := \{v \text{ vertex in } D \mid B_{\frac{\pi}{2}}(v) \cap \text{Im}(\gamma) \neq \emptyset\}$ is nonempty. The vertices in this set are pairwise connected by edges. But, this contradicts the fact that γ can not be a union of two geodesics of length π. Thus, $\text{Im}(\gamma) \subset \text{span}(B) \subset D$. But, $\text{span}(B)$ is a simplex in D as D is flag. Therefore γ is contained in a simplex. However, by assumption γ is a closed local geodesic. We thus arrive at a contradiction and conclude that the statement must hold. $\square$

Using Proposition 4.42 one can refine Gromov's link condition in case of cube complexes. Combining Propositions 4.22 and 4.42 we directly obtain the following theorem.

Theorem 4.43 (Gromov's Link Condition for Cube Complexes) *A finite dimensional cube complex is locally* CAT(0) *if and only if all its vertex links are flag simplicial complexes. In particular a cube complex X is* CAT(0) *if and only if it is simply connected and all its vertex links are flag.*

Remark 4.44 The analog of Theorem 4.43 for infinite dimensional cube complexes is also true and shown by Ian Leary in Appendix B of [Lea13]. Hence, one may drop the finite dimensional assumption on the cube complex and also use this result in a much wider context.

It is not hard to check that the links of vertices in the standard cubing of $\mathbb{R}^n$ are flag simplicial complexes. This implies, by Gromov's link condition that these spaces are CAT(0). Let's see one more example of an application of the link condition.

Example 4.45 Figure 4.13 illustrates how to turn a surface of genus two into a cube complex. The construction is done as follows.

Take a cubical cylinder as shown in the middle of Fig. 4.13. The edges of the left and right-hand side 4-gon of the cylinder are colored and oriented according to the blue arrows. Respecting colors and directions glue the edges of both 4-gons to the corresponding edges in the figure-eight on the left. The resulting object is almost a cube complex.

This space has a unique vertex corresponding to x, two edges corresponding to the edges with single and double blue arrows and four additional edges, which are color coded in yellow, green, blue and red. The link of the unique vertex x has the shape of a complete bipartite graph on six vertices as shown on the right-hand side of Fig. 4.13. In addition there are four 2-cells corresponding to the squares in the cubical cylinder.

---

[1] For details compare Davis-Moussong [DM99] Lemma 2.3.6.

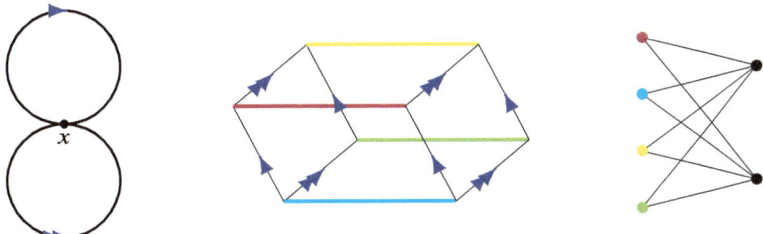

Fig. 4.13 Consider the cube complex which is obtained by gluing the cylinder in the middle to the figure eight on the left by identifying the directed edges with a same arrow. The complete bipartite graph on the right is then the link of the vertex x in this complex. Details provided in Example 4.45

The graph on the right-hand side is a flag simplicial complex. So we almost constructed a CAT(0) cube complex. Except that some of the squares are glued to themselves. However, one can resolve these self-gluings by cubically barycentrically subdividing each original edge and square. Each square of the cubical cylinder is then composed of four squares of half the side length. This results in new vertices in the complex all of whose links are 4-gons and hence flag.

We thus obtain a CAT(0) cube complex, which is topologically a genus two surface.

4.3 Cube Completions

In this section we highlight the close connection between locally non-positive curvature and the 2-skeleta of cube complexes following Haglund and Wise [HW08]. We prove that cube complexes can sometimes be completed to locally CAT(0) cube complexes.

Definition 4.46 (Completable Complexes) A cube complex X is *completable* if there exists a locally CAT(0) cube complex Y and an isomorphism from the two-skeleton of X to the 2-skeleton of Y. We call Y a *completion* of X.

Note that every locally CAT(0) space is trivially completable and serves as its own completion. Our next goal is to prove that every combinatorial map defined on a 2-skeleton extends to the whole complex. The precise statement is given below in Lemma 4.51. We now introduce some additional notation used in the remainder of this subsection.

Notation 4.47 Recall that a k-cube is a set $C = C^k = [0, 1]^k \subset \mathbb{R}^k$. We write $C^{k,2}$ for the 2-skeleton of a k-cube and denote by D_k the subcomplex of C formed by the union of all 2-faces containing the vertex $v_k = (1, 1, \ldots, 1)$ of C. The union of all edges containing v_k is abbreviated by T_k. Moreover, for an edge e containing

v_k denote by σ_e the Euclidean reflection that maps the cube to itself and fixes e. In other words, σ_e is the reflection on the affine hyperplane transversal to e as defined in Definition 5.3.

Definition 4.48 (A k-Corner) A k-*corner* in a cube complex X is a combinatorial map $D_k \to X$. Two k-corners $c_i : D_k \to X, i = 1, 2$, are *adjacent along an edge e* in T_k if $c_2 \circ \sigma_e = c_1$.

Remark 4.49 Note that if there are two k-corners $c_i : D_k \to X, i = 1, 2$ with $c_1(v_k) = c_2(v_k)$ adjacent along the same edge e, then $c_1 = c_2$. See [HW08] Remark 12.3.

Non-positively curved cube complexes can be characterized using k-corners.

Lemma 4.50 *A cube complex X is locally* CAT(0) *if and only if for all $k \geq 3$ any k-corner extends to exactly one k-cube of X, that is, any combinatorial map $d : D_k \to X$ has a unique extension $c : C \to X$ with $c|_{D_k} = d$.*

Proof Suppose X is locally CAT(0). By the definition of a cube complex there is at most one extension c of any k-corner d, which implies uniqueness. Existence follows from the the local CAT(0) property, which implies that links are flag. The converse also follows from Gromov's link condition Theorem 4.43. □

Proposition 4.51 (Extending Combinatorial Maps) *Let X and Y be cube complexes and suppose that Y is locally* CAT(0). *Then every combinatorial map $f : X^{(2)} \to Y$ extends to a unique combinatorial map $X \to Y$.*

Proof Lemma 4.50 implies that any two potential extensions of f have to agree on each cube of X. Hence, if an extension exists it is unique. To show existence it is enough to prove that every combinatorial map $c : C^{k,2} \to Y$ extends to a unique k-cube in Y. Restrictions of the map c define a collection of k-corners, which we denote by $\{c_v\}_{v \in C^{k,2}}$. But then, Lemma 4.50 implies that each corner c_v extends to a unique k-cube $\bar{c}_v : C^k \to Y$. Given two adjacent vertices v and w in C^k the corners c_v and c_w are adjacent along the edge e spanned by v and w. Hence, $c_v \circ \sigma_e = c_w$ and thus also $\bar{c}_v \circ \sigma_e = \bar{c}_w$. The group G_k generated by the reflections σ_e with e an edge of T_k acts transitively on all vertices of the k-cube, compare Example 4.72. This implies that $\bar{c}_v = \bar{c}_{v_k} \circ \sigma_v$, where $v_k = (1, 1, \ldots, 1)$ again and σ_v is the element of G_k mapping v_k to v. This shows that $\bar{c}_{v_k}$ is a k-cube equal to c on each of the corners $\sigma_v(D_k)$ and hence extends c uniquely. □

This proposition has interesting direct consequences.

Corollary 4.52 *Any completable cube complex X combinatorially embeds in a locally* CAT(0) *cube complex $\bar{X}$ by a map that induces an isomorphism on the respective 2-skeleta.*

Proof Let X be a completable cube complex and $i : X^{(2)} \to \bar{X}^{(2)}$ an isomorphism to the two-skeleton of a locally CAT(0) cube complex $\bar{X}$. Then Lemma 4.51 implies that there exists a unique extension $\bar{i} : X \to \bar{X}$ of i. The map i is injective and links of X are simplicial complexes, which implies that also the extension $\bar{i}$ of i must be injective. Hence, the assertion. □

The previous corollary implies that every completable cube complex has a unique CAT(0) completion.

Definition 4.53 (Cube Completion) A *cube completion* of a completable complex is a map $\bar{i}$ as in the proof of Corollary 4.52.

Using the next lemma one can show that cube-completions are in fact unique, compare Exercise 4.73.

Lemma 4.54 *Let X, Y be two completable cube complexes with cube completions $i : X \to \bar{X}$ and $j : Y \to \bar{Y}$. For every combinatorial map $f : X \to Y$ there exists a unique combinatorial map $\bar{f} : \bar{X} \to \bar{Y}$ such that $\bar{f} \circ i = j \circ f$.*

Proof Consider the map $j \circ f \circ (i^{-1})|_{\bar{X}^{(2)}} : \bar{X}^{(2)} \to \bar{Y}$. The space $\bar{X}$ is completable as a locally CAT(0) space. We may thus apply Lemma 4.51 to this map and obtain a unique combinatorial map $g : \bar{X} \to \bar{Y}$ such that $g = j \circ f \circ (i^{-1})|_{\bar{X}^{(2)}}$ on $\bar{X}^{(2)}$ and hence also $g \circ i = j \circ f$ on $X^{(2)}$. But then, the uniqueness of the combinatorial extension, provided by Lemma 4.51, implies that this equality holds on all of X. □

We finish this subsection with the following lemma, which provides a sequence of nested cube complexes and combinatorial embeddings that are close to a cube completion. The proof is taken from [HW08].

Lemma 4.55 *For any finite dimensional cube complex X there exists a cube complex $\bar{X}$ together with a combinatorial embedding $j : X \to \bar{X}$ such that*

1. *the map j restricts to an isomorphism of the the 2-skeleta and*
2. *every combinatorial map $C^{n,2} \to \bar{X}$ extends to a unique n-cube of X.*

Proof Inductively define a sequence of spaces $X = X_2, X_3, \ldots$ together with combinatorial embeddings $i_k : X_k \to X_{k+1}$ such that i_k restricts to an isomorphism of the 2-skeleton and such that every combinatorial map $C^{l,2} \to \bar{X}$ extends to a unique l-cube of X_k for all $l \geq k$. The limit of this system of inclusions exists and has the desired properties.

Suppose the sequence has been constructed for some $k \geq 2$. Now the restriction of every combinatorial map $c : C^{k+1,2} \to X_k$ to the intersection of a codimension one face of C^k with $C^{k,2}$ is given by a combinatorial map $q : C^{k,2} \to X_k$. By the induction hypothesis the map q extends to a map $\bar{q}$ from the union of the codimension one faces of the $(k+1)$-cube to X_k. By the definition of cube complexes such an extension is unique.

Declare two combinatorial maps $f : C^{k,2} \to X_n$ and $g : C^{k,2} \to X_n$ to be equivalent if there is an automorphism ϕ of the $(n+1)$-cube, which satisfies $f \circ \phi = g$. If f, g are equivalent and f extends to the k cube, then so does g.

Pick a representative q_α for each equivalence class of combinatorial maps $C^{n+1,2} \to X_n$ that does not extend to an $(n + 1)$-cube of X_n. Denote its extension to $\partial C^{n+1} \to X_n$. Attach an $(n + 1)$-cube to X_n for each such map and denote the resulting complex by X_{n+1}. By construction $X_n \subset X_{n+1}$ and the two spaces have the same 2-skeleton.

For $l \leq n$ every combinatorial map from the 2-skeleton of the l-cube to X_{n+1} extends uniquely to some l-cube.

Consider a combinatorial map $f : C^{n+1,2} \to X_{n+1}$. If f extends to an $(n+1)$-cube of X_n, then any extension to X_{n+1} has range in X_n (since f does) and hence is unique.

In case f does not extend to an n-cube of X_n it is equivalent to one of the q_α. But then, by construction of X_{n+1} these maps uniquely extend to X_{n+1}. Hence, so does f.

We finish the proof by checking that the gluing of cubes in X_{n+1} satisfies the definition of a cube complex. In particular this means that the links in X_{n+1} are simplicial complexes. The complex X_n is a cube complex by the induction hypothesis. Any $(n+1)$-corner admits an extension, which is unique. This implies that no bad gluings appear and X_{n+1} is again a cube complex. By putting $\bar{X}$ to be the direct limit of the inclusions $X_2 \to X_3 \to X_4 \to \ldots$ we obtain that the inclusion $X \to \bar{X}$ has the desired properties.   □

We refer the reader to Sect. 7.3 and in particular Proposition 7.73 where the notion of a completion is revisited and linked to a specific subclass of cube complexes.

4.4   Right Angled Artin Groups

Artin groups, also known as Artin–Tits groups, are a family of infinite groups defined by presentations of a certain type. The defining relations are all of the form $sts \ldots = tst \ldots$ where both sides have the same length. Artin groups are closely related to Coxeter groups.

Right-angled Artin groups (short RAAGs) are a subclass of Artin groups with defining relations being just commutators. Their presentation can be encoded in a graph. Examples of RAAGs include free groups and free abelian groups of finite rank. RAAGs have decidable word- and conjugacy problem and there exists an explicit construction of a finite $K(\pi, 1)$ space.

In this section we define RAAGs and prove that every such group admits a geometric action on a CAT(0) cube complex—its Salvetti complex.

Definition 4.56 (Right-Angled Artin Groups) Let $\Gamma = (V, E)$ be a finite graph with V the set of vertices and E the set of edges of Γ. A *right angled Artin group* $A(\Gamma)$, short a *RAAG*, is generated by a collection of generators g_v with $v \in V$ where two such generators g_u and g_v commute whenever the vertices u, v are connected by an edge in Γ. That is, $A(\Gamma)$ is given by the following presentation:

$$A(\Gamma) := \langle g_v \text{ for } v \in V \mid [g_u, g_v] = 1 \text{ for all edges } (u, v) \in E \rangle.$$

Note that we may also define infinitely generated RAAGs by means of infinite graphs in the same way.

We will see that RAAGs naturally act on cube complexes. In case the circumference of Γ is greater or equal than four, the presentation 2-complex of the RAAG $A(\Gamma)$ is locally CAT(0), see Example 4.74, and the group acts nicely on the universal cover of this space. However, in general higher dimensional cubes are needed in order to obtain a locally CAT(0) complex from a RAAG. This is achieved by the Salvetti complex introduced later in this section.

Let's have a look at some first examples which illustrate the fact that RAAGs interpolate between free groups and free abelian groups.

Example 4.57 The RAAG defined by a complete graph on n vertices is the free abelian group on n generators. The group $A(\Gamma)$ is isometric to the free group F_n on n generators if Γ consists of n disconnected vertices with no edges. Other examples interpolate between these two extremes. Consider the graphs Γ_1 and Γ_2 pictured below.

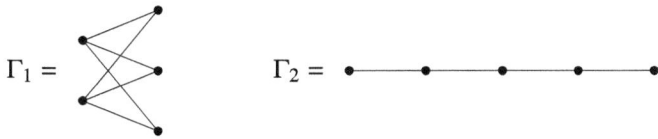

The RAAG defined by the bipartite graph Γ_1 is isomorphic to a direct product $F_2 \times F_3$. The RAAG defined by Γ_2 is isomorphic to the fundamental group of a knot complement. Namely the complement of the figure ⬭⬭⬭⬭ embedded in $\mathbb{S}^3$.

Two right-angled Artin groups $A(\Gamma)$ and $A(\Gamma')$ are abstractly isomorphic if and only if Γ and Γ' are. See Droms [Dro87] for a proof. The obvious question now is, how is the structure of the defining graph is related to algebraic properties of the group. We first study how RAAGs behave with respect to taking sub-graphs of the defining graph.

Definition 4.58 (Induced Subgraph) Let $\Gamma = (V, E)$ be a graph. A graph $\Gamma' = (V', E')$ is a subgraph of Γ if $V' \subset V$ and $E' \subset E$. we say Γ' is *induced* if E' consists of all edges in E having both their vertices in V'. That is, two vertices in V' are connected by an edge in E' if and only if they are connected by an edge in E.

Take an n-gon. Then an edge path consisting of n vertices connected by n_1 edges in a row forms a line which is a subgraph of the n-gon. This subgraph is not induced. However, any line of k vertices with $k < n$ is isometric to an induced subgraph.

Induced subgraphs of defining graphs of Artin groups correspond to injections of smaller groups as subgroups of the Artin group.

Lemma 4.59 *Every morphism* $f : \Gamma_1 \to \Gamma_2$ *of graphs induces a homomorphism* $f^* : A(\Gamma_1) \to A(\Gamma_2)$ *of groups.*

Proof The map f^* is defined by $f^*(v) = f(v)$ for all $v \in V_1 \subset A(\Gamma_1)$ where we view V_i as a subset of the group $A(\Gamma_i)$. One needs to prove that $f^*(ab) =$

$f^*(a) \cdot f^*(b)$ for all a, b in $A(\Gamma_1)$. It is clear that $f^*(\mathbb{1}) = \mathbb{1} \in A(\Gamma_2)$. Since f is a morphism of graphs, for every edge (v, w) in Γ_1 either $f(v) = f(w)$ or $(f(v), f(w))$ is an edge in Γ_2. But then, the generators corresponding to v, w in $A(\Gamma_1)$ commute if and only if the generators $f^*(v)$ and $f^*(w)$ commute in $A(\Gamma_2)$. This implies the assertion.   □

Proposition 4.60 (RAAGs for Induced Subgraphs) *Let Γ be a graph and Γ' an induced subgraph. Let $\iota : \Gamma' \to \Gamma$ denote the inclusion map. Then the induced map $\iota^* : A(\Gamma') \to A(\Gamma)$ on the right-angled Artin groups is an injective homomorphism.*

Proof Define a map $\pi : A(\Gamma) \to A(\Gamma')$ by putting $\pi(v) = v$ for every generator whose corresponding vertex v is in Γ' and by putting $\pi(v) = \mathbb{1}$ whenever v is not in the subgraph Γ'. The homomorphism induced by π has the property that $\pi \circ \iota^* : A(\Gamma') \to A(\Gamma')$ is injective. Thus, ι^* has to be injective.   □

This proposition has several consequences stated below.

Proposition 4.61 (Properties of RAAGs) *Let $\Gamma = (V, E)$ be a finite graph and let $A(\Gamma)$ be the RAAG defined by Γ. Then*

1. *$A(\Gamma)$ is abelian if and only if Γ is complete.*
2. *$A(\Gamma)$ is a direct product of two groups A_1 and A_2 if and only if the defining graph Γ is a join of two graphs Γ_1 and Γ_2. That is, the vertex set V of Γ splits as a disjoint union of two subsets V_1 and V_2 and between every pair of vertices $v_1 \in V_1$ $v_2 \in V_2$ there exists an edge in E.*
3. *$A(\Gamma)$ is the free product of two groups A_1 and A_2 if and only if the defining graph Γ has two distinct and non-empty connected components.*

Proof To see the first item observe that two generators of $A(\Gamma)$ commute if and only if the vertices in Γ corresponding to them are connected by an edge. By Proposition 4.60 every pair of vertices not connected by an edge generate a free subgroup of rank two. So if the defining graph is not complete the group can not be abelian. For the second item we first deduce from Proposition 4.60 that every pair of induced subgraphs generate two subgroups. The assertion then follows from the first item, since in a join of groups A_1 and A_2 every element (or generator) of A_1 has to commute with every element (or generator) of A_2. Similar arguments imply the third item.   □

Note that not all subgroups of a RAAG are defined by an induced subgraph. In particular not all subgroups of RAAGs are finitely generated. Examples for non-finitely generated subgroups can be obtained by looking at kernels of projections of RAAGs onto $\mathbb{Z}$. Subgroups not defined by subgraphs include fundamental groups of surfaces.

In the general setting of finitely generated groups it is not easy to characterize subgroups of a randomly chosen group. Nor is it in general easy to embed RAAGs into other classes of groups. Davis and Januszkiewicz [DJ00] proved that every right-angled Artin group is isomorphic to a finite index subgroup of a right-angled Coxeter group. A right-angled Coxeter group is a as defined in Definition 6.1, all whose defining pairwise orders m_{ij} are either two or infinity. Let's have a look at a small example.

Example 4.62 Let A be the free abelian group $\mathbb{Z}^2$ generated by two commuting generators v_0 and v_1. Let W be the right-angled Coxeter group given by the following presentation:

$$W = \langle a_0, b_0, a_1, b_1 \mid a_i^2, b_i^2, a_i b_j = b_j a_i \text{ for all } i, j \rangle.$$

One has

$$W \cong (\mathbb{Z}/2\mathbb{Z} * \mathbb{Z}/2\mathbb{Z}) \times (\mathbb{Z}/2\mathbb{Z} * \mathbb{Z}/2\mathbb{Z}).$$

Define a map $f : A \to W$ via $v_0 \mapsto a_0 a_1$ and $v_1 \mapsto b_0 b_1$. In order to verify that this is an isomorphism onto a subgroup in W one has to show that $f(r) = \mathbb{1}$ for all relations r in A. There is only one relation in A, namely $v_0 v_1 = v_1 v_0$. We have that $f(v_0 v_1 v_0^{-1} v_1^{-1}) = a_0 a_1 b_0 b_1 (a_0 a_1)^{-1} (b_0 b_1)^{-1}$. Since $a_i^{-1} = a_i$ and $b_i^{-1} = b_i$ this reduces to $f(v_0 v_1 v_0^{-1} v_1^{-1}) = \mathbb{1}$, which completes the proof.

RAAGs admit a natural action on a nice CAT(0) cube complex. This complex arises as the universal cover of the Salvetti complex.

Definition 4.63 (Salvetti Complex) Let $\Gamma = (V, E)$ be a finite graph. The *Salvetti complex* $S(\Gamma)$ is the complex constructed as follows:

The vertex set of $S(\Gamma)$ contains a single element x_0. There exists one edge e_v for every vertex $v \in V$, which we label with color v and glue both its ends to the unique vertex x_0. Hence, the 1-skeleton $S^{(1)}$ of the Salvetti complex $S(\Gamma)$ is a rose with k petals, where $k = |V|$ and where the edge-labels are in bijection with V.

Inductively construct the k-skeleton $S^{(k)}$ of $S(\Gamma)$ by gluing k-cubes for every k-clique, that is, every complete graph on k vertices of Γ, as follows. Suppose $k = 2$. For every edge $\{u, v\}$ in E take a square and label two parallel faces by u and the two remaining faces by v. Glue this 2-cube to the rose by matching edges with a same label. More formally, $S^{(2)}$ is the quotient of the union of $S^{(1)}$ with a collection of squares, one for each edge, by the gluings of faces according to colors.

For $k > 2$ suppose the $(k - 1)$-skeleton has been constructed. For every k-clique in Γ take a k-cube and inductively color its l-faces by subsets of the vertices of Γ corresponding to sub-cliques of size l. Start by coloring the vertices of the cube— each with a different color, which is a vertex of Γ. The coloring is done in such a way that the set coloring an l-face is contained in the color of every l'-face containing it. The k-skeleton is then obtained as the quotient of the union of $S^{(k-1)}$ with the collection of all these k-cubes. Two faces are identified in the quotient whenever they carry the same color.

The dimension of the Salvetti complex is the size of the largest clique in Γ.

By construction, the RAAG $A(\Gamma)$ is isomorphic to the fundamental group $\pi_1(S(\Gamma))$ of the Salvetti complex. See Example 4.65 for the construction of the Salvetti complex of $\mathbb{Z}^3$.

Remark 4.64 (Barycentric Subdivision of the Salvetti Complex) It is worth emphasizing that a Salvetti complex $S(\Gamma)$ is never a cube complex in the sense defined in 4.24. The problem is that in $S(\Gamma)$ all cubes have self-gluings, unless Γ (and thus $S(\Gamma)$) consists of a single vertex.

However, one can turn a Salvetti complex into a cube complex by barycentrically subdividing its cubes. To obtain the *barycentric subdivision* sd(C) of a cube C place a new vertex in the barycenter of every face F of C. Connect such a barycenter to the centers of the faces of F that are of codimension one in F. For example, a square will be subdivided into four squares by cutting the square trough the midpoints of its parallel sides. The four smaller squares share the barycenter of the original square. Each cube of dimension k is replaces by 2^k cubes of the same dimension. In case two cubes are glued along common faces the smaller cubes inherit the gluing. More formally, the faces of the barycentric subdivision correspond to chains of non-empty faces in the face poset of the original complex.

The barycentric subdivision $\check{S}(\Gamma) = \mathrm{sd}(S(\Gamma))$ of the Salvetti complex is a cube complex with extra good structure of its set of hyperplanes. More on this *special* behavior in Sect. 7.3.

Example 4.65 Let Γ be a triangle graph consisting of three vertices x, y, and z pairwise connected by an edge. Then Γ is the defining graph of the free abelian group $\mathbb{Z}^3$ on three generators written as a right-angled Artin group. Figure 4.14 shows how to construct the Salvetti complex of $\mathbb{Z}^3$. The 1-skeleton of the complex is a rose with three petals, shown on the very left of Fig. 4.14. The three generators x, y and z are subject to the relations $a = xyx^{-1}y^{-1}$, $b = xzx^{-1}z^{-1}$ and $c = yzy^{-1}z^{-1}$. Any pair of vertices is connected by an edge. Hence, each of these edges adds a

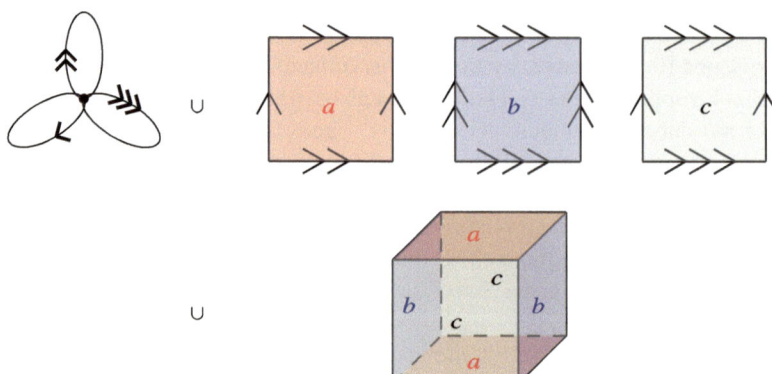

Fig. 4.14 This figure shows how to construct the Salvetti complex of the RAAG $\mathbb{Z}^3$. See Example 4.65 for details

square to the complex. The three squares are shown in blue, red and gray and are labeled with a, b, and c. In the quotient these squares each yield a 2-torus. The largest clique is the triangle on x, y, z itself. This clique yields a 3-cube whose edges are labeled with the three generators and the sides by pairs of generators. Hence each of the sides looks like one of the three 2-cubes labeled a, b, c. We have labeled the sides of the 3 cubes with the same letters and colors in the figure. The Salvetti complex is a 3-torus obtained as the quotient of the rose, the 2-cubes and the 3-cube shown in Fig. 4.14 by identifying edges with a same label and faces of a same color.

Charney and Davis [CD95] proved the following result. The cube complex $R(\Gamma)$ constructed in the proof of this theorem is a cubical subdivision of the universal cover of the Salvetti complex of $A(\Gamma)$. It corresponds to the standard presentation complex of $A(\Gamma)$, while $A(\Gamma)$ is isomorphic to the fundamental group $\pi_1(S(\Gamma))$.

Theorem 4.66 (Charney-Davis Theorem) *For every finite, simplicial graph Γ there exists a locally CAT(0) cube complex $R(\Gamma)$ on which the group $A(\Gamma)$ acts by isometries. The complex $R(\Gamma)$ is a subdivision of the universal cover of the Salvetti complex $S(\Gamma)$. The action of G on $R(\Gamma)$ is free, proper and cocompact.*

Proof Consider the barycentric subdivision $\hat{S}(\Gamma)$ of the Salvetti complex as defined in Remark 4.64. The universal cover of $\hat{S}(\Gamma)$ is called $R(\Gamma)$. By construction $A(\Gamma) = \pi_1(S(\Gamma))$. Hence, the RAAG acts on $R(\Gamma)$ by deck transformations.

It remains to prove that $R(\Gamma)$ is CAT(0), i.e. all links in $R(\Gamma)$ are flag. We may check this condition on the subdivision $\hat{S}(\Gamma)$ of $S(\Gamma)$. The only vertex where one has to prove something is the unique vertex of $\hat{S}(\Gamma)$ inherited from $S(\Gamma)$. This vertex x_0 has the same link in both complexes. One may hence compute its link in $S(\Gamma)$.

Every vertex in $\mathrm{lk}_{S(\Gamma)}(x_0)$ corresponds to a generator in $A(\Gamma)$. Any two such vertices are joined by an edge in $\mathrm{lk}_{S(\Gamma)}(x_0)$ whenever there is a square glued onto the corresponding two petals in $S(\Gamma)$. This is the case if and only if the generators corresponding to the petals commute in $A(\Gamma)$. More generally, every k-simplex in $\mathrm{lk}_{S(\Gamma)}(x_0)$ corresponds to a clique of pairwise commuting generators in $A(\Gamma)$. For each such group there is a k-torus, which fills the simplex. So there are no empty simplices and the link is flag. This implies that $R(\Gamma)$ is locally CAT(0).   □

General Artin groups are much harder to study then right-angled ones. Apart from very few cases and selected sub-classes many basic properties remain mysterious. In the general case the following basic questions remain open: solving the word and conjugacy problems, computing the center or the torsion, the $K(\pi, 1)$ conjecture and the question whether or not these groups act geometrically on a CAT(0) cube complex. Cubulations have been obtained in several special cases. See, for example, the work of Huang, Jankiewicz and Przytycki [HJP16] or by Haettel [Hae21, Hae22] and the references provided therein.

4.5 Exercises

Exercise 4.67 Describe all the isometry classes of cube complexes that arise from quotients of the cubical barycentric subdivision of a single square.

Exercise 4.68 Show that a right-angled spherical complex is a CAT(1) space if and only if it is a local CAT(1) space and there are no locally geodesic closed cycles of length $< 2\pi$.

Exercise 4.69 Let C be an M_κ^n polyhedral cell. Prove that

1. C is the convex hull of its vertices.
2. For every set P of points in the model space with the property that $C = \mathrm{conv}(P)$ there exists a minimal subset P' of P such that $C = \mathrm{conv}(P')$ and such that $P' = C^{(0)}$ is the set of vertices of C.

I.e. prove items 4 and 5 of Theorem 4.6.

For a hint see the footnote.[2]

Exercise 4.70 Let K be an M_κ polyhedral complex and $x, y \in K$, such that $d(x, y) < \epsilon(x)$. Then every cell $C \subseteq K$ containing y also contains x and one has that $d(x, y) = d_C(x, y)$. Here d is the pseudo-metric on K and d_C the natural metric on the cell C.

Exercise 4.71 Show that a closed edge-path in a CAT(0) cube complex has even length.

Exercise 4.72 We use notation as in Notation 4.47. Write the reflections σ_e as matrices and prove explicitly, using coordinate descriptions of the vertices that the action of G_k is transitive on the set of vertices of the k-cube C. Moreover, prove that every vertex stabilizer is trivial.

Exercise 4.73 Prove that any two cube completions of a completable cube complex are isomorphic. That is, for two given cube completions $i_j : X \to \bar{X}_j$ with $j = 1, 2$ show that there exists a unique isomorphism $f : X_1 \to X_2$ such that $f \circ i_1 = i_2$.

Exercise 4.74 Look up the definition of a presentation 2-complex in [Lö17] and prove that this complex is CAT(0) for a RAAG defined by a graph whose girth is at least four (which means that there are no cycles of length three in the graph).

---

[2] Use induction for the proofs.

Fig. 4.15 This figure shows
an example of a VH-complex

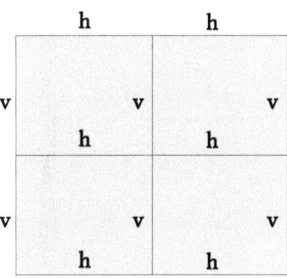

Exercise 4.75 A *VH-complex* is a marked cube complex of dimension 2 such that every edge is either labeled with a v (for vertical) or with h (for horizontal) and the attaching map of every 2-cell alternates between v and h (which means that parallel edges in a single square share the same label). A simple example is shown in Fig. 4.15. Prove that VH-complexes are locally CAT(0).

Chapter 5
Hyperplanes and Half-Spaces

A fundamental tool in the study of cube complexes are hyperplanes and the half-spaces they define. Hyperplanes are defined by means of parallel edges inside 2-cubes and carry themselves the structure of a CAT(0) cube complexes of a smaller dimension. Each hyperplane in a CAT(0) cube complex divides the complex into two disjoint half-spaces.

Surprisingly the combinatorics of the relative position of the hyperplanes and half-spaces completely determines the cube complex. The set of half-spaces has a natural poset structure whose combinatorics has a huge impact on the groups acting geometrically on the cube complex. From the poset of half-spaces one can reconstruct the whole space and also define a boundary.

In the first section we introduce hyperplanes and half-spaces. Sageev's cubulation method constructs a cube complex from an abstract half-space system. We discuss this method in the second section. In the third section we explain Roller duality, that is the connection between a cube complex, the associated half-space system and the complex constructed from this system by Sageev's method. Doing so we are also able to introduce the Roller boundary of a cube complex.

5.1 Hyperplanes

In this section we introduce hyperplanes, prove that they are CAT(0) cube complexes and investigate their structure. The definition of hyperplanes and half-spaces relies on the notion of parallel edges.

Definition 5.1 (Parallel Edges) Two edges e, e' in a cube complex X are *elementary parallel*,[1] denoted by $e \sim e'$, if e is opposite e' in some 2-dimensional cube in

---

[1] Sometimes such a pair of edges is called *square equivalent* .

© The Author(s), under exclusive license to Springer Nature Switzerland AG 2023
P. Schwer, *CAT(0) Cube Complexes*, Lecture Notes in Mathematics 2324,
https://doi.org/10.1007/978-3-031-43622-2_5

X. We extend $\sim$ to an equivalence relation on all edges in X. That is, two edges are *parallel* if there exists $k \in \mathbb{N}$ and a sequence $e_0 = e, e_1, \ldots, e_k = e'$ of edges such that e_i and e_{i-1} are elementary parallel. An equivalence class $[e]$ of parallel edges is called *wall* of X.

Every wall in a cube complex X gives rise to a sub-structure, called hyperplane, one dimension lower whose cubes sit inside cubes of X perpendicular to the edges comprising the wall. In case X is CAT(0) these hyperplanes are cube complexes themselves.

Definition 5.2 (Midcube) A *midcube* M_i, for some $i \in \{1, \ldots, n\}$, of a cube $C = [0, 1]^n$ is a subset of the cube of the following form:

$$M_i := \left\{ x \in C \mid x_i = \frac{1}{2} \right\}.$$

We omit the index and write midcube M if there exists some i for which $M_i = M$. Midcubes of cubes in a cube complex are the images of midcubes in the cubes before applying the gluing maps. A midcube M is *perpendicular* to a wall $[e]$ (or simply to e) if the intersection of M with the 1-skeleton of X consists of midpoints of edges in $[e]$. We write $M \pitchfork e$.

A hyperplane is the collection of all midcubes perpendicular to a fixed wall.

Definition 5.3 (Hyperplanes) Let X be a cube complex. A *hyperplane* $H \subset X$ is the union of all midcubes M perpendicular to the parallelism class of a fixed edge e.

$$H(e) := \bigcup_{M \pitchfork [e]} M.$$

The *support* $N(H)$ of a hyperplane H is the union of all cubes C in X intersecting H in a midcube.

Alternatively, one can define the hyperplane $H(e)$ determined by an edge e to be the abstract complex of midcubes defined by

$$H(e) = \bigsqcup_{M \pitchfork [e]} M / \sim$$

where the cubes are glued according to the rules in X. In this case the hyperplane is not considered to be a subcomplex of X. As the two definitions yield isomorphic complexes we do not distinguish between them.

There are two possibilities of the intersection of a hyperplane H with a cube in a CAT(0) cube complex. For any given cube C the intersection $C \cap H$ is either empty or a single midcube. Here are some first examples of hyperplanes in CAT(0) cube complexes.

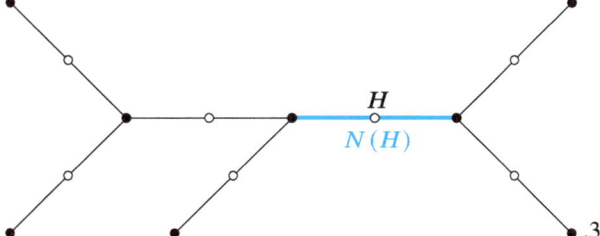

Fig. 5.1 Trees are CAT(0) cube complexes whose hyperplanes consist of single points

Example 5.4 Cube complexes of dimension one are metric realizations of graphs. Recall that a graph is CAT(0) if and only if it has no cycles and is hence a tree. An example of a one-dimensional CAT(0) cube complex is shown in Fig. 5.1. The eight vertices of that graph are drawn as black bold dots. The midpoints of edges are marked with empty circles.

No pair of edges in a graph is contained in a common square, so no two edges are elementary parallel. Hence, walls in a graph consist of single edges and hyperplanes are single vertices represented by the midpoints of the edges. The support of any hyperplane H in the tree is a single edge. One such pair of a hyperplane and it its support is shown in Fig. 5.1. The edge corresponding to the hyperplane (midpoint) labeled H is drawn with a bold light blue edge.

More interesting examples of CAT(0) cube complexes and their hyperplanes are provided in the following example.

Example 5.5 Figures 5.2 and 5.3 highlight two examples of CAT(0) cube complexes and their hyperplanes.

1. Euclidean 2-space $\mathbb{R}^2$ with its standard cubing has two parallel classes of edges. The walls and hyperplanes align with the horizontal and vertical axes as shown in Fig. 5.2. Two examples, labeled H and H', of a vertical and a horizontal hyperplane are shown in this figure. The support $N(H)$ of the hyperplane H is shaded in light blue.

Fig. 5.2 $\mathbb{R}^2$ with the standard cubing is a CAT(0) cube complex. See Example 5.5 for details

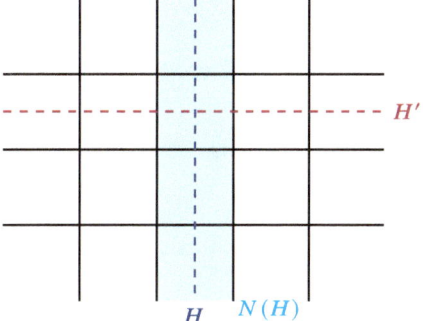

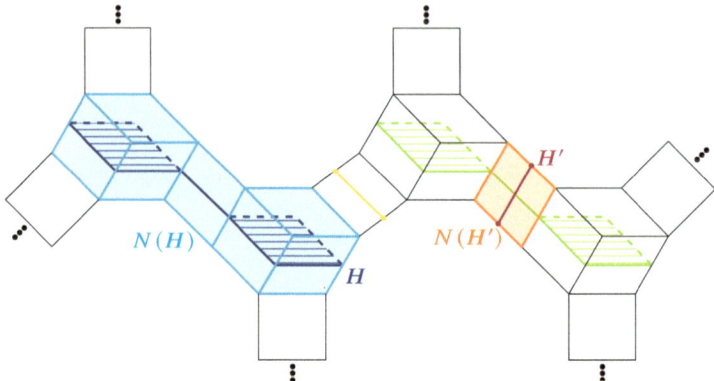

Fig. 5.3 The CAT(0) cubulation of $PGL_2(\mathbb{Z})$ with some of its hyperplanes. See Examples 5.5 and 6.35 for details

2. Another example of a CAT(0) cube complex is shown in Fig. 5.3. This complex consists of two- and three-dimensional cubes arranged in a very symmetric way to form an infinite complex. We learn in Example 6.35 how to construct this complex from $PGL_2(\mathbb{Z})$. This group naturally acts on this complex. There are two different isometry classes of hyperplanes. The hyperplanes in one class consist of a single 1-dimensional midcube. An example of such a hyperplane is shown in red and labeled H'. The support of H' is shaded in orange. The other types of hyperplanes consists of an edge with 2-dimensional cubes attached at both its vertices. Examples are shown in blue and green in Fig. 5.3. One of them is labeled H. The support of H is shaded in light blue. Both types of hyperplanes are finite cube complexes although the ambient complex is itself infinite.

The hyperplanes of CAT(0) cube complexes are well behaved and satisfy a couple of nice properties, which we now prove. Recall, by Proposition 3.17, that locally isometric embeddings into CAT(0) spaces are isometric embeddings.

Proposition 5.6 (Hyperplanes in CAT(0) **Cube Complexes)** *Every hyperplane of a finite dimensional* CAT(0) *cube complex X is closed and convex in X. The support $N(H)$ is a convex subcomplex of X isometric to $H \times [0, 1]$.*

The proof of this proposition is taken from seminar notes by Pascal Rolli [Rol12] where it is attributed to Ruth Charney.

Proof We start with the proof of the first assertion. Let H be a hyperplane defined by $[e]$. Put $X_0 := N(H)$ and let $\tilde{X}_0$ be the universal cover of X_0 with the induced cube structure. The class of e lifts to a wall $[\tilde{e}]$ in $\tilde{X}_0$. Observe that in $\tilde{X}_0$ each maximal cube contains exactly one midcube perpendicular to $\tilde{e}$ since otherwise $\tilde{X}_0$ would not be simply connected. Hence there exists an isometry $f : \tilde{X}_0 \to \tilde{X}_0$, which is uniquely defined by the property that it maps a maximal cube to its reflected image along the midcube perpendicular to $\tilde{e}$. The fixed point set Fix(f) then equals

$\tilde{H}$, where $\tilde{H}$ is the hyperplane defined by $\tilde{e}$. Proposition 3.14 now implies that $\tilde{H}$ is closed and convex in $\tilde{X}_0$. From Proposition 3.17 we obtain that $g : \tilde{X}_0 \to X_0 \hookrightarrow X$, with g being the concatenation of the covering map from $\tilde{X}_0$ to X_0 followed by the local isometric embedding of X_0 into X, is in fact an isometric embedding itself. Since $g(\tilde{H}) = H$ we may deduce that H is closed and convex.

The second assertion can easily be deduced from the first and its proof using that for a midcube M in C one has $C \cong M \times [0, 1]$. $\square$

It is not hard to verify the statement of Proposition 5.6 for the examples provided by Figs. 5.2 and 5.3. Observe that in these examples the hyperplanes are in fact CAT(0) cube complexes themselves, which divide the cube complex into two disjoint pieces. This is a property not all cube complexes share but which holds true in the CAT(0) setting.

Proposition 5.7 (Hyperplanes are CAT(0)**)** *A hyperplane H in a finite dimensional CAT(0) cube complex X is itself a CAT(0) cube complex. For each cube C in X the intersection $H \cap C$ is either empty or a midcube of C. The complement $X \setminus H$ has precisely two connected components.*

Proof Observe first that the intersection of H with a cube is a unique midcube by Proposition 5.6. Hence, to show that H is a cube complex it is enough to convince oneself that a given wall $[e]$ induces the following equivalence relation on midcubes: two midcubes $M_i \subset C_i$, i=1,2, are equivalent, denoted by $M_1 \sim M_2$, if there exists a common face F of C_1 and C_2 such that $M_1 \cap F = M_2 \cap F$. We write $[M]_\sim$ for the equivalence class of a midcube M under the equivalence relation induced by $\sim$. The CAT(0) property then follows from Proposition 5.6 as convex subsets of CAT(0) spaces are again CAT(0).

We now prove that the complement has two components. Let H be a hyperplane defined by the wall $[e]$ and let m be the midpoint of the edge e as shown in Fig. 5.4. An edge e is closed and convex in X hence there is, by Proposition 3.11, a projection $\pi_e : X \to X$ such that $d(x, \pi_e(x)) = \inf_{y \in e} d(x, y)$.

We claim that $\pi_e^{-1}(\{m\}) = H$ and $\pi_e(H) = \{m\}$, where π_H is the projection onto H from Proposition 3.11.

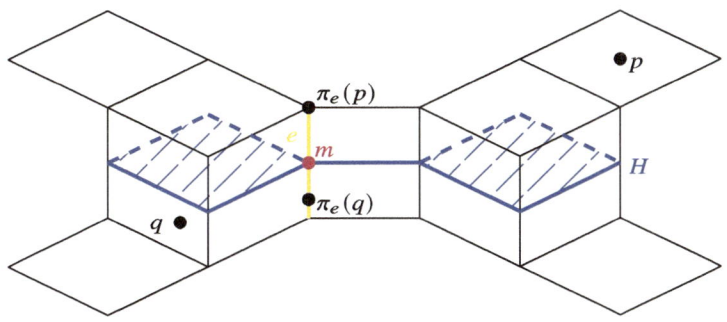

Fig. 5.4 Illustration of a step in the proof of Proposition 5.7

Choose $x \in H$, then $\pi_H(\pi_e(x)) = m$ and

$$d(x, m) = d(\pi_H(x), \pi_H(\pi_e(x))) \le d(x, \pi_e(x)).$$

The point $\pi_e(x)$ is the unique point in e having minimal distance to x. Hence, the claim.

Suppose that $\pi_e(x) = m$ for some $x \in X$. We prove by contradiction that $x \in H$. Assume the contrary. Then the geodesic γ connecting x and m shares only m with H. Moreover γ and e do not form a right angle at m. Otherwise $\gamma \cap H$ would contain more than one point. But then, γ may be shortened to a curve $\tilde{\gamma}$ perpendicular to e and connecting x with e, which contradicts Proposition 3.11. We thus obtain that $\pi_e(X \setminus H) = e \setminus \{m\}$. Recall that X is a geodesic space and hence each point in $X \setminus H$ can be connected by a geodesic to precisely one of the two pieces of $e \setminus \{m\}$. □

Sageev provided a different proof of Proposition 5.7: Studying links of vertices in H one can show that H is locally CAT(0). Simply connectedness can be obtained using disk-diagram techniques. Compare [Sag95].

The assertion of Proposition 5.7 allows us to define half-spaces. We show in Sect. 5.2 that from these hyperplanes a CAT(0) cube complex may be reconstructed. By the definition below $X = h \cup H \cup h^c$.

Definition 5.8 (Half-Spaces)   Let H be a hyperplane in a CAT(0) cube complex X. The connected components of $X \setminus H$ are called *(open) half-spaces* of X. We write h and h^c for the two half-spaces determined by H.

We finish this section by examining intersections of hyperplanes. Compare this result with Helly's theorem for CAT(0) cube complexes stated as Theorem 7.18 in this book.

Proposition 5.9 (Intersections of Hyperplanes) *Let X be a CAT(0) cube complex and let $H_1, \ldots, H_m$ be hyperplanes in X such that $H_i \cap H_j \ne \emptyset$ for all $i \ne j$. Then*

$$\bigcap_{i=1}^{m} H_i \ne \emptyset.$$

If $\dim X = n < \infty$ each family $\{H_i\}_{i \in I}$ with $H_i \cap H_j \ne \emptyset$ for all $i \ne j \in I$ contains at most n elements.

Proof The proof is by induction on m. We consider $m = 3$ as a smallest case. Let H_1, H_2 and H_3 be hyperplanes. Suppose further that $H_1 \cap H_2 \cap H_3 = \emptyset$. Let γ be a geodesic connecting $H_3 \cap H_1$ with $H_2 \cap H_3$ inside the hyperplane H_3 and let x and y denote its endpoints as shown in Fig. 5.5.

Let now z be a point in $H_1 \cap H_2$ and consider geodesics segments $\mathrm{seg}(z, x) \subset H_1$ and $\mathrm{seg}(z, y) \subset H_2$. Then consider the triangle Δ spanned by these geodesics in the vertices x, y, z. The three hyperplanes intersect transversely therefore the angles in

Fig. 5.5 Illustration of a step
in the proof of Proposition 5.9

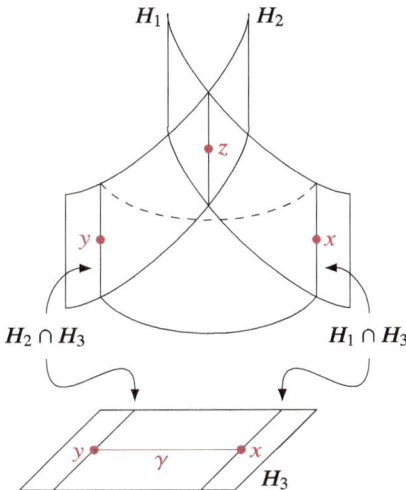

Δ at x and y are $\frac{\pi}{2}$. This is a contradiction as such a triangle does not exist in $\mathbb{R}^2$ but we have assumed the space to be CAT(0).

For the induction step suppose $m > 3$ and that the assertion is true for $m - 1$. Let $H_1, H_2, \ldots, H_m$ be a collection of hyperplanes. Hyperplane H_m is in particular a cube complex itself and intersection $H_i \cap H_m$ are hyperplanes in H_m. By the induction hypothesis the collection $\{H_i \cap H_m\}_{i=1,\ldots,m-1}$ has a non-empty intersection. Thus, $\bigcap_{i=1}^{m} H_i \neq \emptyset$.

It remains to prove that if the dimension of X is n, then there exist only finitely many elements in such a collection. Let $\{H_i\}_{i \in I}$ be a set of hyperplanes with pairwise nonempty intersection. Then $\bigcap_{i \in I} H_i \neq \emptyset$, that is, there exist $|I|$-many midcubes, which intersect nontrivially. Each n-cube has only n such midcubes. Thus, $|I| \leq n = \dim(X)$. □

5.2 Half-Space Systems

The main goal of this section is to understand how cube complexes are determined by their collection of half-spaces or hyperplanes. Based on the behavior of hyperplanes in examples we have encountered, we first define an abstract notion of half-space systems and show how one can construct a CAT(0) cube complex from them.

Recall from Proposition 5.7 that the complement of a hyperplane has two connected components. These components are the half-spaces defined in Definition 5.8. We may introduce an order relation on the set of all such half-spaces in a CAT(0) cube complex X by saying one half-space is smaller than the other if it is contained in the other half-space. The next definition is an abstraction of this concept capturing some of the key properties. We provide examples later in the section.

Definition 5.10 (Half-Space System)   Consider a triple $(\mathcal{H}, \preceq, \star)$ with $\mathcal{H}$ a set, $\preceq$ a partial order on $\mathcal{H}$ and $\star$ an order reversing involution $\star : \mathcal{H} \to \mathcal{H} : h \mapsto h^\star$. We define three conditions:

1. (Complementation) $h \neq h^\star$, the elements h and $h^\star$ are incomparable, and $g \preceq h$ implies $h^\star \preceq g^\star$.
2. (Finite interval condition) for $h_1, h_2 \in H$ there exist only finitely many $k \in \mathcal{H}$ with $h_1 \preceq k \preceq h_2$.
3. (Nesting condition) for distinct $h, k \in \mathcal{H}$ at most one of the following is true: $h \preceq k, h \preceq k^\star, h^\star \preceq k$ and $h^\star \preceq k^\star$.

A triple satisfying only the three properties listed in the first condition, named complementation, is called a *pocset*[2]

A *half-space system* is a triple HSS $= (\mathcal{H}, \preceq, \star)$ satisfying all three conditions, namely complementation, the finite interval condition and the nesting condition. In this case elements $h \in \mathcal{H}$ are referred to as *half-spaces*. If none of the conditions in item 3 is satisfied for a fixed pair h, k we say that h and k are *transversal*.

From a half-space system we may define abstract notions of hyperplanes as follows.

Definition 5.11 (Hyperplanes from Half-Space Systems, Width)   Let the triple $(\mathcal{H}, \preceq, \star)$ be a half-space system. Defining an equivalence relation by putting $h \sim h^\star$ we obtain a set $\bar{\mathcal{H}} := \mathcal{H}/\sim$ of *hyperplanes* $\bar{h} \in \bar{\mathcal{H}}$. We identify an equivalence class $\bar{h}$ with the defining set $\{h, h^\star\}$. The *boundary map* $\delta : \mathcal{H} \to \bar{\mathcal{H}}$ assigns to a half-space h its hyperplane $\bar{h} = \{h, h^\star\}$.

The *width* of a half-space system is the cardinality of a maximal subset $\mathcal{K} \in \mathcal{H}$ such that all pairs $\bar{h}, \bar{k} \in \mathcal{K}$ are transversal.

The following example illustrates why we may think of $\bar{h}$ as an actual wall separating the half-spaces h and $h^\star$ in some kind of ambient space.

Example 5.12 (Half-Space System from a Cube Complex)   Let X be the standard cubing of the Euclidean plane as pictured in Fig. 5.6. Put $\mathcal{H}$ to be the set of all half-spaces in X determined by hyperplanes as defined in Definition 5.8. The order relation $\preceq$ is defined by putting $h \preceq h'$ whenever $h \subset h'$ as subsets of X. The order reversing map $\star$ swaps the two half-spaces h and h^c of a wall $\bar{h}$. With this we obtain a half-space system in the sense of Definition 5.10. Figure 5.6 shows a selection of half-spaces and associated walls, for example, the half-spaces $k, k^\star$ and its wall $\bar{k}$ are drawn in blue. In the same figure, the half-spaces h and k are examples of transversal half-spaces, and the pair $h^\star, h'$ is nested, as $h^\star$ contains h'.

---

[2] Pocset is short for poset with complementation. See [Rol98].

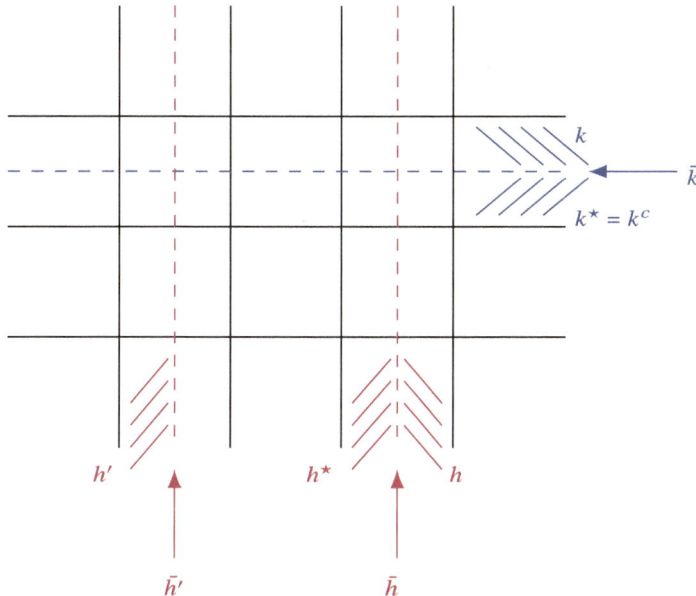

Fig. 5.6 The half-spaces h, h' and k in the standard cubing of $\mathbb{R}^2$ are such that h and h' are nested, and h and k are transversal

In fact, every CAT(0) cube complex gives rise to a half-space system in the way described in Example 5.12. See the next theorem.

Theorem 5.13 (CAT(0) **Cube Complexes Yield Half-Space Systems**) *Let X be a finite dimensional,* CAT(0) *cube complex. Let $\mathcal{H}$ be the set of half-spaces in X (as defined in Definition 5.8) ordered by inclusion, so $h \preceq k$ if $h \subset k$ as subspaces of X. Define a map $\star : H \to H$ by sending a half-space h to its complement h^c, that is, $h^\star = h^c$. Then $(\mathcal{H}, \preceq, \star)$ is a half-space system.*

By construction such a triple $(\mathcal{H}, \preceq, \star)$ is a pocset. The remaining properties, the finite interval and nesting condition take a bit more work to verify. A proof can, for example, be found as Theorem 10.3 in Roller's Habilitationsschrift [Rol98]. Roller also mentions that Gerasimov [Ger98] provides an alternative proof of this result using different methods.

Definition 5.14 (Half-Space System of X) For a finite dimensional CAT(0) cube complex X the *half-space system of X* is the half-space system described in Theorem 5.13. We denote it as $(\mathcal{H}(X), \preceq, \star)$.

The converse is also true. One can construct a cube complex from a half-space system. The remainder of this section is devoted to describing this construction. The original method is due to Sageev [Sag95]. But, other authors have also contributed to the presented formulation. See, for example, Chatterji and Niblo [CN05] or Nica [Nic04]. The main statement is the following.

Theorem 5.15 (Cubulation of Half-Space Systems) *A given half-space system*
$(\mathcal{H}, \preceq, \star)$ *of finite width defines a finite dimensional cube complex* $X(\mathcal{H})$. *Every
connected component of* $X(\mathcal{H})$ *is* CAT(0) *and maximal cubes are in a one-to-one
correspondence with maximal families of pairwise transversal hyperplanes. The
dimension of a maximal cube equals the number of hyperplanes in such a family.*

Before we are able to give a proof on page 95 we need to see how one constructs
a cube complex from a half-space system. In the formal definition of $X(\mathcal{H})$ vertices
are abstract maps. However, we first provide a heuristic description and explain
some examples.

Main Idea Concerning Vertices
In the standard cubing of the plane, see, for example, Fig. 5.6, half-spaces are
defined by vertical or horizontal lines. In this complex for every half-space h any
given vertex v is either contained in h or in its complement $h^\star$ as shown on the left-
hand side of in Fig. 5.7. It may never happen that $v \in h \cap h^\star$. This is automatically
satisfied if the half-space system is constructed from a cube complex by means of
hyperplanes in the complex.

Moreover, a vertex in a cube complex is uniquely determined by the set of half-
spaces containing it. This should be guaranteed by the construction as well. We also
would like to have the following property satisfied by the constructed cube complex:
Whenever $v \in h$ and we have $h \subset k$ then v should be also contained in k as shown
on the right in Fig. 5.7.

We now provide the formal definition.

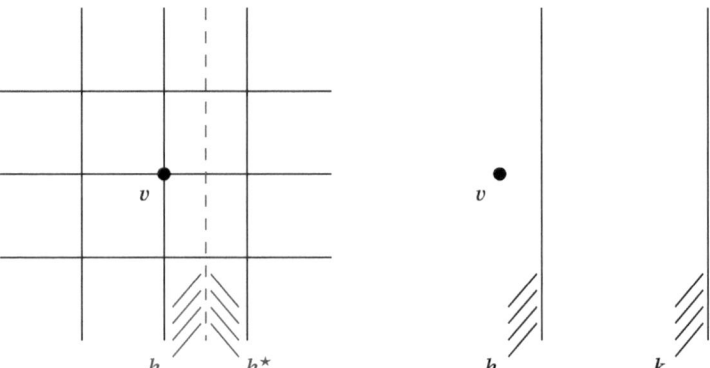

Fig. 5.7 The half-spaces h and k contain v, the half-space $h^\star$ does not contain the vertex v

Fig. 5.8 Definition 5.16
excludes assignments by
vertex-maps as shown in the
picture as they would imply
that $v(\bar{h}) \preceq v(\bar{k})^\star$

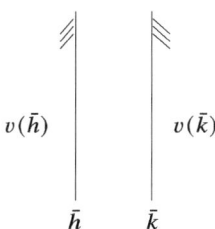

$v(\bar{h})$ $v(\bar{k})$

$\bar{h}$ $\bar{k}$

Definition 5.16 (The 1-Skeleton of the Cubulation)   Fix a half-space system
$(\mathcal{H}, \preceq, \star)$. The 1-skeleton of the *cube complex* $X(\mathcal{H})$ associated with the given half-space system is defined as follows:

1. The set of *vertices* $V(\mathcal{H})$ is defined by

$$V(\mathcal{H}) = \left\{ v : \bar{\mathcal{H}} \to \mathcal{H} \mid v(\bar{h}) \in \bar{h} \text{ and } v(\bar{h}) \npreceq (v(\bar{k}))^\star \text{ for all } \bar{k}, \bar{h} \in \bar{\mathcal{H}} \right\}.$$

Recall that $\bar{h} = \{h, h^\star\}$ and hence $v(\bar{h}) \in \bar{h}$ means that v maps the class $\bar{h}$ to one of its defining elements. We write $v \in h$ in case $v(\bar{h}) = h$ and say that v is *contained in h*.
2. The set of *edges* $E \subset V \times V$ consists of all pairs of vertices $\{v, w\}$ such that

$$\left| \{ \bar{h} \in \bar{\mathcal{H}} \mid v(\bar{h}) \neq w(\bar{h}) \} \right| = 1.$$

The map $v : \bar{\mathcal{H}} \to \mathcal{H}$ can be thought of as assigning to each hyperplane $\bar{h}$ the half-space in $\bar{h} = \{h, h^\star\}$ that contains v. So in particular inconsistent choices of half-spaces for parallel walls, as shown in Fig. 5.8, do not occur.

We continue defining the higher dimensional cubes, soon. But, first let us introduce additional properties of the vertices and half-spaces. The neighbors of a vertex can be characterized using minimal half-spaces.

Definition 5.17 (Minimal Half-Space)   A half-space h is *minimal* with respect to a vertex $v \in V(\mathcal{H})$ if $v \in h$ and there does not exist $k \in \mathcal{H} \setminus \{h\}$ with $v \in k$ and $k \prec h$.

Examples of (non-)minimal half-spaces are shown in Fig. 5.9 where h is minimal with respect to v, but not minimal with respect to the vertex w. In the same figure the half-space k is minimal with respect to w, but not with respect to the vertex v. We now describe neighbors in terms of altered maps.

The map $v_{\bar{h}}$ in Definition 5.18 swaps the half-space h for its opposite half-space $h^\star$. Thus, if $v_{\bar{h}}$ is a vertex as defined in Definition 5.16, then v and $v_{\bar{h}}$ are adjacent vertices in $X(\mathcal{H})$.

Fig. 5.9 The half-space k is
minimal with respect to v.
The half-space h is minimal
with respect to w but not with
respect to v

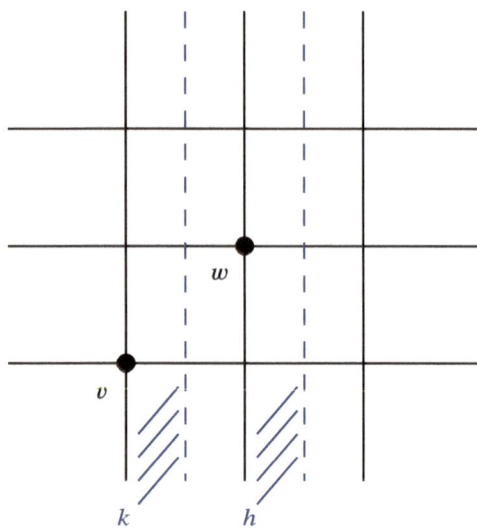

Definition 5.18 (Neighbors) Let v be a vertex in $X(\mathcal{H})$, let $\bar{h} \in \bar{\mathcal{H}}$ and define $v_{\bar{h}} : \bar{\mathcal{H}} \to \mathcal{H}$ by

$$v_{\bar{h}}(\bar{k}) = \begin{cases} v(\bar{k})^{\star} & \text{if } \bar{k} = \bar{h} \\ v(\bar{k}) & \text{else.} \end{cases} \tag{5.1}$$

If $v_{\bar{h}}$ is itself a vertex, then we call $v_{\bar{h}}$ the *neighbor* of v *along* $\bar{h}$. We also say that $\bar{h}$ separates $v_{\bar{h}}$ and v.

The next lemma characterizes those maps defined in Definition 5.18, which are vertices.

Lemma 5.19 *Let* $v \in V(\mathcal{H}), \bar{h} \in \bar{\mathcal{H}}$ *and let* $v_{\bar{h}}$ *be as in Definition 5.18. Then*

1. *$v_{\bar{h}}$ is a vertex if and only if h is minimal with respect to v. And*
2. *the neighbors of a vertex v are in one-to one correspondence with the half-spaces h that are minimal with respect to v. More precisely with the hyperplanes $\bar{h} = \{h, h^{\star}\}$ such that h or $h^{\star}$ is minimal with respect to v.*

Proof We first prove item 1. Let $v_{\bar{h}}$ be a vertex. Then for all $\bar{k} \in \bar{H}$ the half-space $v_{\bar{h}}(\bar{k})$ is not contained in $(v_{\bar{h}}(\bar{h}))^{\star}$.

Suppose for a contradiction that h is not minimal with respect to v. There exists then $k \neq h$ with $v \in k$ and $k \preceq h$. By Definition 5.18 we then have that $v_{\bar{h}}(\bar{k}) = k$ and

$$v_{\bar{h}}(\bar{k}) = k \preceq h = (v_{\bar{h}}(\bar{h}))^{\star}.$$

This contradicts the fact that $v_{\bar{h}}$ is a vertex.

To prove the converse let h be minimal with respect to v and suppose that $v_{\bar{h}}$ is not a vertex. Then one of the two following alternatives is satisfied:

- $v_{\bar{h}}(\bar{h}) \preceq (v_{\bar{h}}(\bar{k}))^{\star}$ for some $\bar{k} \in \bar{\mathcal{H}}$
- $v_{\bar{h}}(\bar{k}) \preceq (v_{\bar{h}}(\bar{h}))^{\star}$ for some $\bar{k} \in \bar{\mathcal{H}}$.

Note that it can not happen that h is not involved since v is a vertex that has only been altered in the image of h.

Suppose without loss of generality that $v_{\bar{h}}(\bar{h}) \preceq (v_{\bar{h}}(\bar{k}))^{\star}$ for some $\bar{k} \in \bar{\mathcal{H}}$ holds. Then

$$h^{\star} = (v(\bar{h}))^{\star} = v_{\bar{h}}(\bar{h}) \preceq (v_{\bar{h}}(\bar{k}))^{\star} = (v(\bar{k}))^{\star}.$$

Using the fact that * is orientation reversing we may conclude that $v \in v(\bar{k}) \preceq h$, which contradicts the assumption that h was minimal.

Item 1 and the definition of vertices and half-spaces imply that item 2 is true. $\square$

One is now in the position to define the higher dimensional cube. As for vertices let us first get a better understanding of the underlying ideas.

Heuristic for the Definition of n-Cubes

Let h and k be transversal hyperplanes and assume they are minimal with respect to a vertex v. Changing v on one of these hyperplanes results in a neighbor of v. Changing v on both h and k at the same time should again yield a vertex, which is adjacent to both neighbors of v. In other words a pair of transversal hyperplanes should correspond to a 2-cube in the complex. This is exactly what happens in our running example $\mathbb{R}^2$ as is illustrated in Fig. 5.10.

Following up on Definition 5.16 we continue with the definition of the n-cubes in $X(H)$.

Definition 5.20 (n-Cubes of the Cubulation) Let $(\mathcal{H}, \preceq, \star)$ be a half-space system and denote by $X^1(\mathcal{H})$ the 1-skeleton of a cube complex as defined in Definition 5.16. Inductively define the n-skeleton $X^n(\mathcal{H})$ of $X(\mathcal{H})$ as follows:

3. Glue an n-cube to every subcomplex Y of $X^{n-1}(\mathcal{H})$ that is isomorphic to the $(n-1)$-skeleton of an n-cube C^n.

Fig. 5.10 Every 2-cube corresponds to a pair of transversal hyperplanes

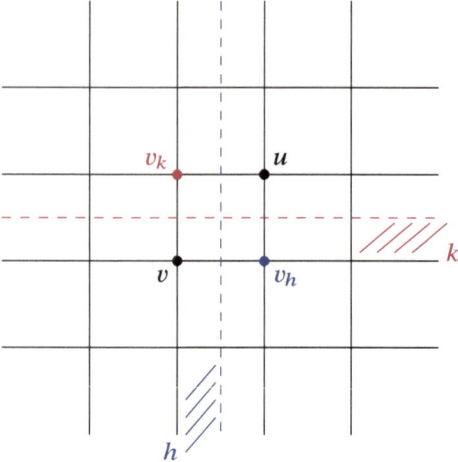

Fig. 5.11 The neighbors of a vertex v are determined by minimal hyperplanes. See Lemma 5.21

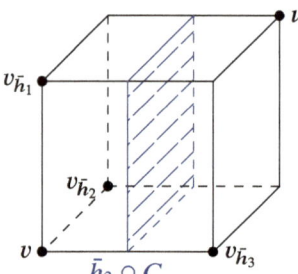

Every empty n-cube in $X(\mathcal{H})$ is filled in by this construction. One keeps adding cubes as long as there are empty skeleta of higher-dimensional cubes. It is, a priori, not clear whether or not this process ends in finitely many steps and results in a finite dimensional complex.

We now characterize neighbors of vertices using hyperplanes. Figure 5.11 shows the neighbors $v_{\bar{h}_i}$ of a vertex v in a single 3-cube. Each of these neighbors corresponds to a unique hyperplane in the cube separating $v_{\bar{h}_i}$ from v. One of these hyperplanes is shown in blue.

A general characterization is given by the following lemma.

Lemma 5.21 *Let C be an n-cube in $X(\mathcal{H})$ and v a vertex of C. The neighbors of v in C are of the form $v_{\bar{h}_i}$ for hyperplanes $\bar{h}_i$, $i = 1 \ldots, n$ where $v_{\bar{h}_i}$ is as defined in Definition 5.18. Let u be the vertex in C diagonally opposite v, then u may be obtained from v by simultaneously switching out h_i by $h_i^\star$ for all i in the definition of v. That is,*

$$v_{\bar{h}}(\bar{k}) = \begin{cases} v(\bar{k})^\star & \text{if } \bar{k} = \bar{h}_i \text{ for some } i = 1, \ldots, n \\ v(\bar{k}) & \text{else.} \end{cases} \tag{5.2}$$

Proof The proof is by induction on n. For $n = 1$ there is nothing to prove. For $n = 2$ there are four vertices in the cube in total. The vertex v, its neighbors v_1, v_2 and the diagonally opposite vertex u. We know that v labels the half-spaces in H in such a way that v corresponds to the map $\bar{h} \mapsto h$ for all $\bar{h} \in \bar{\mathcal{H}}$. By slight abuse of notation ($\bar{\mathcal{H}}$ might not be countable) we may think of this map as listing all the half-spaces chosen by v, that is, for a suitable naming of the elements of $\bar{\mathcal{H}}$ and $\mathcal{H}$ we have that

$$v \hat{=} (h_1, h_2, h_3, \ldots), \quad v_1 \hat{=} (h_1^\star, h_2, h_3, \ldots) \text{ and } v_2 \hat{=} (h_1, h_2^\star, h_3, \ldots).$$

The map that represents vertex four needs to differ from v_1 and v_2 in precisely one position each. Hence, there is no choice other than putting

$$u \hat{=} (h_1^\star, h_2^\star, h_3, \ldots).$$

Suppose now that we have shown the assertion for $(n - 1)$ and $(n - 2)$ and let v be a vertex in an n-cube C having neighbors $v_i := v_{\bar{h}_i}$ for $i = 1, \ldots, n$. The vertex diagonally opposite v in C is again denoted by u. We write σ_i for the codimension one face of C containing all vertices v_j, $j = 1, \ldots, n$ except for vertex v_i and denote by w_i the vertex diagonally opposite v in σ_i. Compare Fig. 5.12. By the induction hypothesis the vertex w_i was obtained from v by simultaneously switching out the half-spaces h_j for all $j \neq i$. That is,

$$w_i(\bar{k}) = \begin{cases} v(\bar{k})^\star & \text{if } \bar{k} = \bar{h}_j \text{ for some } j \neq i \\ v(\bar{k}) & \text{else.} \end{cases}$$

Consider the vertex v' obtained from v by simultaneously switching out the hyperplanes $h_1, \ldots, h_{n-2}$. This vertex is diagonally opposite v in the (n-2)-subcube of C containing the vertices $v_1, \ldots, v_{n-2}$. Now v', w_n and w_{n-1} are three vertices of a 2-dimensional subcube of C. Since u is a common neighbor of w_n and w_{n-1} it has to be the fourth vertex in this cube and is obtained from v' by simultaneously switching out the half-spaces h_{n-1} and h_n. This proves the lemma. $\quad\square$

The next lemma characterizes cubes in terms of sets of hyperplanes. Compare also Fig. 5.13 where some of the vertices inside a 3-cube are labeled according to that construction.

Lemma 5.22 *Let v be a vertex in $X(\mathcal{H})$ and let $S \subset \bar{\mathcal{H}}$ be a collection of hyperplanes with $|S| = n$. For every subset T of S define a vertex w_T by*

$$w_T(\bar{k}) = \begin{cases} v(\bar{k})^\star & \text{if } \bar{k} \in T \\ v(\bar{k}) & \text{if } \bar{k} \notin T. \end{cases}$$

Fig. 5.12 Illustration of the
induction step in the proof of
Lemma 5.21

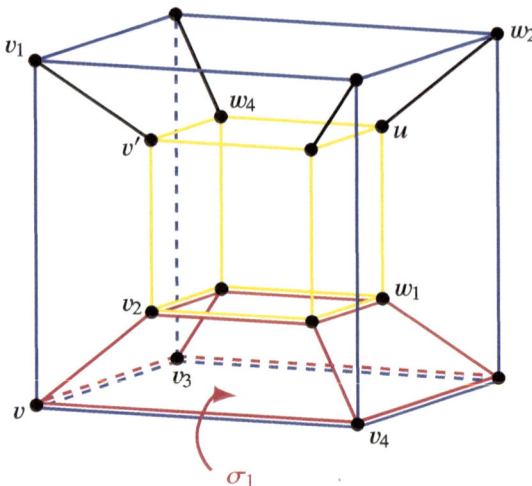

Fig. 5.13 Every set of n
pairwise transitive
hyperplanes spans an n-cube
as shown in Lemma 5.22

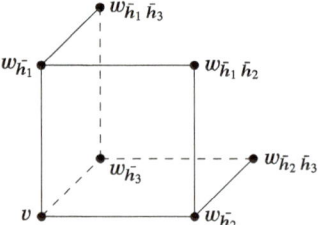

*The set of vertices $\{w_T \mid T \subseteq S\}$ spans an n-cube if and only if the following two
properties hold:*

1. the half-space $v(\bar{h})$ is minimal with respect to v for all $\bar{h}$ in S and
2. the image $v(\bar{h})$ is transversal to $v(\bar{k})$ for all $\bar{h}$ and $\bar{k}$ in S.

Proof Put $h := v(\bar{h})$ and $k := v(\bar{k})$. Suppose first that the given vertices span an n-
cube. Minimality follows from Lemma 5.19 item 1 as $v(\bar{h})$ is minimal with respect
to v for all $\bar{h} \in S$ (here $|T| = 1$). In C the vertex v is adjacent to all vertices w_T
with $|T| = 1$.

We now prove transversality, i.e., item 2. Choose $T = \{\bar{h}, \bar{k}\} \subset S$ and write
$w_{\bar{k}}$, respectively $w_{\bar{h}}$, instead of $W_{T'}$ for the vertices defined by the subsets T' of T
containing a single hyperplane.

Lemma 5.21 implies that v, $w_{\bar{h}}$, $w_{\bar{k}}$ and w_T span a 2-cube. Further h is minimal
with respect to v and $w_{\bar{k}}$ and similarly k is minimal with respect to v and $w_{\bar{h}}$. From
the Definition 5.17 of minimality we obtain that

$$\nexists\, h' \in \mathcal{H} \text{ such that either } v \in h', \text{ or else } w_{\bar{k}} \in h' \text{ and } h' \preceq h, \tag{5.3}$$

and the same statement with the roles of h and k switched. Putting $h' = k$ in Eq. (5.3) we get

$$v(\bar{h}) = h \not\preceq k \Longleftrightarrow h^\star \not\preceq k^\star$$

and

$$w_{\bar{k}}(\bar{k}) = v(\bar{k})^\star = k^\star \not\preceq h \Longleftrightarrow h^\star \not\preceq k.$$

Similar statements with roles of h and k switched rule out the remaining cases. Therefore h and k are transversal.

To prove the converse suppose that items 1 and 2 are satisfied. We first show by induction on $n := |T|$ that w_T is a vertex for all $T \subset S$.

If $n = 1$ the claim follows from Lemma 5.19 since $v(\bar{h}) = h$ is minimal with respect to v for all $\bar{h} \in S$.

Suppose the statement is true for $|T| = n - 1$. For a vertex w_T and $\bar{h} \in S \setminus T$ we need to show that $h := v(\bar{h})$ is minimal with respect to w_T. One may then apply Lemma 5.19 to obtain the assertion for $|T| = n$.

We know that h is minimal with respect to v. Hence, there is no $\bar{k} \in \mathcal{H}$ such that $v(\bar{k}) = k$ and $k \preceq h$. For all $k \in \mathcal{H} \setminus T$ we have that $v(\bar{k}) = w_T(\bar{k})$ and hence

$$\not\exists \bar{k} \in \mathcal{H} \setminus T \text{ such that } w_T(\bar{k}) = k \text{ and } k \preceq h. \tag{5.4}$$

Further by the definition of w_T we know for all $k \in T$ that $(v(\bar{k}))^\star = w_T(\bar{k})$.

But, h is transversal to $k := v(\bar{k})$ for all $\bar{k} \in T$ and hence $h \not\preceq k$, $h \not\preceq k^\star$, $h^\star \not\preceq k$ and $h^\star \not\preceq k^\star$. From this we deduce

$$\not\exists \bar{k} \in T \text{ such that } w_T(\bar{k}) = k^\star \text{ and } k^\star \preceq h. \tag{5.5}$$

Combining Eqs. (5.4) and (5.5) we may deduce that h is minimal with respect to w_T and hence, by Lemma 5.19, the map w_T is in fact a vertex.

It is easy to see that these vertices span the 1-skeleton of an n-cube. And, by construction of $X(\mathcal{H})$, the n-cube itself has to exist. □

We are now ready to prove the main theorem of this section.

Proof of the Cubulation Theorem 5.15 From Lemma 5.22 we obtain that the n-cubes in $X(\mathcal{H})$ are in one-to-one correspondence with the families of n pairwise transversal hyperplanes in H. It hence remains to prove that $X = X(\mathcal{H})$ is locally CAT(0) and each connected component is simply connected.

Consider a vertex $v : \bar{\mathcal{H}} \to \mathcal{H}$ in X. There are no bigons in the link $\mathrm{lk}_X(v)$, since otherwise there would exist pockets in X, compare Fig. 5.14. But, such pockets can not exist since the intersection of a pair of cubes is a face of both.

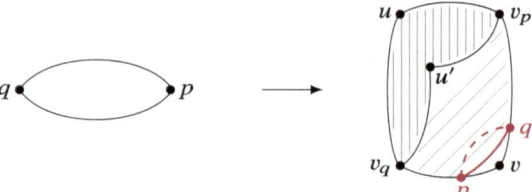

Fig. 5.14 The existence of bigons in links violates the uniqueness of a square determined by transversal hyperplanes

Suppose the link $lk_X(v)$ of v contains the $(n-1)$-skeleton of an n-simplex. Then there are n hyperplanes $\bar{h}_1, \ldots, \bar{h}_n$ such that each $\bar{h}_i$ defines a new vertex v_i adjacent to v. Each pair $\bar{h}_i, \bar{h}_j$ spans a square Q_{ij} (the corresponding edge exists in the link, hence Q_{ij} exists). Thus, conclude by Lemma 5.22 applied to Q_{ij} that h_i is transversal to h_j for al $i \neq j$. Applying Lemma 5.22 to $h_1, \ldots, h_n$ we see that there exists an n-cube having v as a vertex and the Q_{ij} as faces. Hence, the $(n-1)$-skeleton is the boundary of an n-simplex and thus X is locally CAT(0) by Proposition 4.42.

Let Y be a connected component of X. We need to prove that Y is simply connected. The proof of the following two observations is left as an exercise:

- Closed edge-paths in the 1-skeleton $Y^{(1)}$ of Y without backtracking do contain an even number of edges.
- Write $d_1(v, w)$ for the number of hyperplanes separating v and w. This defines a metric on the set of vertices of X and induces a metric on $Y^{(1)}$.

Let v_0 be a vertex in Y and l a closed path in $Y^{(1)}$ containing v_0. Let further v be a vertex on l having maximal distance to v_0 in $Y^{(1)}$ with respect to d_1. There is possibly more than one such vertex. Choose neighbors a, b of v such that v differs from a, respectively b, in precisely one hyperplane $\bar{h}_a$, respectively $\bar{h}_b$, and such that $v(\bar{h}_x) = h_x$ for both $x = a$ and $x = b$. Compare Fig. 5.15.

Claim 1 $\bar{h}_a$ and $\bar{h}_b$ are transversal.

In order to show this claim we have to verify three inequalities: First $h_a \not\leq h_b \not\leq h_a$, then $h_a \not\leq h_b^\star$ and finally $h_a^\star \not\leq h_b$. By construction h_a and h_b are minimal with respect to v. Hence, there does not exist any k with $v \in k$ and $k \preceq h_a$ respectively $k \preceq h_b$. In particular we have that $v(\bar{h}_a) = h_a \not\leq h_b$ and $v(\bar{h}_b) = h_b \not\leq h_a$.

The half-space $h_b^\star$ is minimal with respect to b. Hence, there does not exist a hyperplane $\bar{k}$ with $b(\bar{k}) = k$ and $k \preceq h_b^\star$ and we arrive at the fact that

$$b(\bar{h}_a) = h_a \not\leq h_b^\star.$$

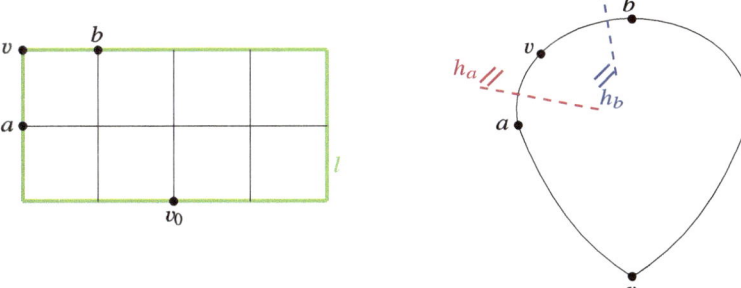

Fig. 5.15 Closed paths in the proof of Theorem 5.15

To prove the last inequality, that is, $h_a^\star \not\leq h_b$, we first show a second claim.

Claim 2 $v_0(\bar{h}_a) = h_a^\star$ and $v_0(\bar{h}_b) = h_b^\star$.

We prove $v_0(\bar{h}_a) = h_a^\star$ by contradiction. The proof of the second claim is along the same lines and left to the reader. Suppose $v_0(\bar{h}_a) = h_a$. The vertices v and v_0 then agree on $\bar{h}_a$ and we have that $\bar{h}_a \in \{\bar{h} \mid v(\bar{h}) \neq v_0(\bar{h})\} := S$. The distance between v and v_0 is $d_1(v, v_0) = |S|$ and

$$a(\bar{h}) = \begin{cases} v(\bar{h}) & \text{if } \bar{h} \neq \bar{h}_a \\ v(\bar{h})^\star & \text{if } \bar{h} = \bar{h}_a. \end{cases}$$

Therefore $\{\bar{h} \mid v_0(\bar{h}) \neq a(\bar{h})\} = S \cup \{\bar{h}_a\}$ and $d_1(v_0, a) = d_1(v_0, v) + 1$, which contradicts the choice of v. Hence, Claim 2.

From Claim 2 we deduce that $h_a^\star = v_0(\bar{h}_a) \not\leq v_0(\bar{h}_b)^\star = h_b^{\star\star} = h_b$, which implies Claim 1. We have thus shown that h_a and h_b are transversal and define a 2-cube in Y having vertices v, a, b and v_{ab}, where v_{ab} is given by

$$v_{ab}(\bar{h}) = \begin{cases} v(\bar{h})^\star & \text{if } \bar{h} = \bar{h}_a \text{ or } \bar{h} = \bar{h}_b \\ v(\bar{h}) & \text{else .} \end{cases}$$

By construction and Claim 2 it follows that $d(v_0, v_{ab}) = d(v_0, v) - 2$.

Define a new loop l' as illustrated in Fig. 5.16 by replacing the two edges connecting a, v, b by the path a, v_{ab}, b and possibly deleting any newly obtained backtracks. Then, since $d(v_0, v_{ab}) < d(v_0, v)$ we obtain by this procedure a homotopy contracting l to v_0. Hence, the assertion. □

We close the section with a last look at our running example.

Example 5.23 Consider the standard cubing X of the Euclidean plane. The edges of this cube complex are shown in black in Fig. 5.17. The set of its walls $\mathcal{H}$ consists of two parallel classes of walls. One class is shown in red, the other class in blue in the same figure. The walls $\bar{h}$ define half-spaces $h, h^\star$ in $\mathcal{H}$, which we have indicated

Fig. 5.16 Defining a
replacement l' of the original
loop—see proof of
Theorem 5.15

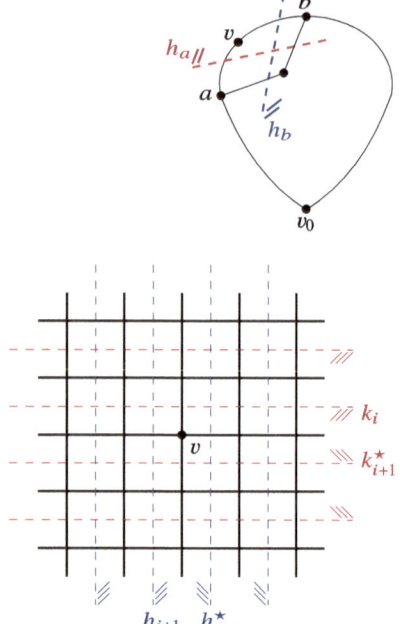

Fig. 5.17 Sageev's
cubulation method gives back
the original complex. The
vertex v corresponds to a map
that assigns to every
hyperplane the half-space
containing v. The correct
choices are, for example, the
half-spaces $h_{i+1}, h_{i+2}^\star \ldots$

by the dashed lines. Every vertex in X corresponds to a map $v : \bar{\mathcal{H}} \to \mathcal{H}$ as
in Definition 5.16 by picking half-spaces containing it for all walls in $\bar{\mathcal{H}}$. This is
illustrated in Fig. 5.17.

The cubulation construction not only describes the original vertices and cubes
but also produces points at infinity of the cube complex one started with. These
boundary points correspond to infinite nested sequences of half-spaces. There are
four lines corresponding to infinite sequences in only one of the two parallel classes
of walls. In case all blue half-spaces are chosen such that they form one nested
infinite family, e.g. all of them pointed to the right, a compatible non-nested choice
for the red hyperplanes produces a vertex in the line to the right at infinity. The
yellow vertex in Fig. 5.19 on Page 103 is, for example, obtained in such a way.

The four points that appear as single vertices correspond to the four ways to orient
the two infinite sequences of nested hyperplanes. These four points are shown on the
left-hand side of Fig. 5.19.

In the next subsection we examine the structure of these additional points and
lines at infinity, and describe in more detail how to get back the original complex
when replacing a cube complex X by its half-space system and then cubulating the
system.

5.3 Roller Duality

Start with a cube complex X, construct its half-space system and from it a cube complex using cubulation of half-space systems, as in Theorem 5.15. Then the resulting complex is in many cases larger than the complex X one started with. We now aim to describe how to recover the original complex inside the cubulation of the half-space system and how to use the cubulation procedure to define the Roller boundary of a cube complex.

Definition 5.24 (Terminating-Chain Vertex) Let v be an element of the vertex-set $V(\mathcal{H})$ of a cubulation of some half-space system $(\mathcal{H}, \preceq, \star)$. Then v is a *terminating chain vertex*, or short *TC-vertex*, if any descending chain in v terminates, that is, every infinite collection of half-spaces $\{h_i := v(\bar{h}_i)\}_{i \in I}$ such that $h_{i+1} \preceq h_i$ for all i eventually stabilizes. Let $X^0(\mathcal{H})$ denote the span of the TC-vertices in $X(\mathcal{H})$.

The next lemma shows that the subspace $X^0(\mathcal{H})$ is a connected component of $X(\mathcal{H})$.

Lemma 5.25 *Let $(\mathcal{H}, \preceq, \star)$ be a half-space system of finite width. Then any two vertices v_x, v_y in $X^0(\mathcal{H})$ are connected by an edge-path in the 1-skeleton. That is, there exists a set S of half-spaces on which the two sections v_x and v_y differ.*

The proof is left as Exercise 5.45. Here is a hint: Consider vertices u, $v \in X^0(\mathcal{H})$. One needs to show that there exist finitely many half-spaces $\bar{h}$ in $\overline{\mathcal{H}}$ such that $v(\bar{h}) \neq u(\bar{h})$. This is enough to prove the lemma as otherwise (using finite width) one can construct a descending, non-terminating chain.

Proposition 5.26 (Terminating Chains Correspond to Vertices) *Let X be a finite dimensional* CAT(0) *cube complex and $(\mathcal{H}(X), \preceq, \star)$ the half-space system defined by X. Then the span $X^0(\mathcal{H})$ of the TC-vertices is isometric to X, i.e.*

$$X^0(\mathcal{H}) \cong X.$$

Moreover, the TC-vertices are in bijection with the vertices in X and any of these vertices is of the form $v_x(\bar{h}) = h \ni x$, where h is a half-space and x a vertex in X.

Proof Put $Y = X^0(\mathcal{H})$. Let x be a vertex in X. Define a vertex by putting $v_x(\bar{h}) = h$ where h is the half-space of $\bar{h}$ which contains x. This vertex terminates as every descending chain terminates in a half-space minimal for x.

This implies that there is a map $\phi^0 : X^0 \to Y^0$ which send a vertex of X to a vertex of Y. This map is injective as any two vertices in X are separated by at least one half-space and therefore define different vertices.

Two adjacent vertices v, w in X connected by an edge e differ in exactly the half-space $\bar{h} := H(e)$ and hence determine an edge in Y. This allows us to extend ϕ^0 to a simplicial map ϕ^1, which maps the 1-skeleton of X into the 1-skeleton of Y.

It is not hard to see that the argument also applies for n-cubes in X and that we therefore obtain an isometric embedding $\phi : X \hookrightarrow Y$.

It remains to prove that ϕ is surjective. Let v be a vertex in Y and let $v_x \in \phi(X^0)$. By Lemma 5.25 there exists a set S on which v and v_x differ and an edge-path in the 1-skeleton of Y that starts in v_x and ends in v. Enumerate the edges in this path in order starting with $\bar{h}_1$ and ending with $\bar{h}_n$. Then, by construction, $S \supset \{\bar{h}_1, \ldots, \bar{h}_n\}$ and $\bar{h}_1$ is minimal with respect to v_x.

The hyperplane $\bar{h}_1$ determines a unique edge e containing v_x. Denote by ω the vertex of e different from v_x in Y. The section ω is a TC-vertex as we have $\omega(\bar{h}) = v_x(\bar{h})$ in case $\bar{h} \neq \bar{h}_1$ and $\omega(\bar{h}) = v_x(\bar{h})^*$ in case $\bar{h} = \bar{h}_1$. This implies that $\omega \in \phi(X)$. Inductively we can find vertices in X by altering the section v_x corresponding to x in the first i hyperplanes $\bar{h}_1, \ldots, \bar{h}_i$ of the sequence given by the edge-path in the 1-skeleton above. Thus, for any v in Y there exists $u \in X$ such that $\phi(u) = v$.

$\square$

There is a duality in the procedures of taking half-space systems and applying Sageev's cubulation construction, which was observed by Roller [Rol98].

Theorem 5.27 (Roller Duality) *Given a half-space system $(\mathcal{K}, \preceq, \star)$ of finite width, then*

$$(\mathcal{H}(X^0(\mathcal{K})), \preceq, \star) = (\mathcal{K}, \preceq, \star).$$

On the other hand, for a finite dimensional CAT(0) *cube complex Y one has*

$$X^0(\mathcal{H}(Y)) = Y.$$

Proof The second assertion stated here is in fact just Proposition 5.26. To see that $\mathcal{H}(X^0(\mathcal{K})) = \mathcal{K}$ argue as follows. Every half-space $k \in \mathcal{K}$ determines a set of vertices $v \in k$ with $v \in X^0(\mathcal{K})$ for which k is minimal. These vertices are contained in edges e whose midpoint is contained in $\bar{k}$. Therefore $\mathcal{H}(X^0(\mathcal{K})) \ni H(e) = k \in K$. And thus $\mathcal{K} \hookrightarrow \mathcal{H}(X^0(\mathcal{K}))$. Let further h be an element of $\mathcal{H}(X^0(\mathcal{K}))$, it is then not hard to check that h has to come from some $k \in \mathcal{K}$. The fact that $\mathcal{H}(X^0(\mathcal{K}))$ is indeed a half-space system follows from Lemma 5.25, which requires $\mathcal{K}$ to be of finite width.

$\square$

Remark 5.28 One can prove that $X(\mathcal{H})$, with $\mathcal{H}$ the set of half-spaces of a cube complex Y, is a compactification of Y with the Roller boundary as defined in Definition 5.34. For details see [Rol98]. In addition an action of a group G on a cube complex Y can be extended to a certain nice subset of the Roller boundary. See [NS13].

Having now characterized cube complexes via half-space systems the question arises whether it is possible to characterize cube subcomplexes of cube complexes via sub-sets of the half-space systems. This is indeed possible.

Definition 5.29 (Subsystem of a Half-Space System) A *subsystem* of a half-space system (also *sub-HSS*) is a subset $\mathcal{K} \subset \mathcal{H}$ (inclusion compatible with the ordering), which is closed with respect to $\star$.

The 1-skeleton of a cube complex is a (metric) graph. With all edges having length one we obtain a natural metric on the 1-skeleton by Definition 2.6. We denote this metric d_1. It is the same as the induced length metric obtained from the restriction of the metric on the whole cube complex to the 1-skeleton. Recall from Definition 5.24 that $X°(\mathcal{H})$ denotes the span of the TC-vertices.

Lemma 5.30 *Let $(\mathcal{K}, \preceq, \star)$ be a subsystem of a half-space system $(\mathcal{H}, \preceq, \star)$. Then there exists a natural projection*

$$p : X(\mathcal{H}) \longrightarrow X(\mathcal{K})$$

that maps $X°(\mathcal{H})$ to $X°(\mathcal{K})$ and does not increase distances with respect to the metric d_1 on the 1-skeleton.

Proof We start by defining the map p on the vertices. Let $v : \bar{\mathcal{H}} \to \mathcal{H}$ be a vertex in X. Put $p(v) := v|_{\bar{\mathcal{K}}}$. It is clear that $p(v)$ is a vertex in Y. Observe further that the number of hyperplanes on which two vertices v and μ differ is always less than or equal to the number of hyperplanes on which $p(v)$ and $p(\mu)$ differ. Hence, the projection map does not increase distances. By Lemma 5.25 the space $X°(\mathcal{K})$ is connected. □

Geometrically one may think of the projection p introduced in the previous proof as follows: Suppose $h \in \mathcal{H} \setminus \mathcal{K}$, then the support of $\bar{h}$ in $X°(\mathcal{H})$ is isomorphic to $I \times \bar{h}$. The map p collapses I in case $h \notin \mathcal{K}$.

Product Decomposition of Cube Complexes
There is another nice property we may look at, given Roller duality. Namely the decomposition of a cube complex as a product of cube complexes.

Theorem 5.31 (Product Decomposition of Half-Space Systems) *Let X be a finite dimensional $\mathrm{CAT}(0)$ cube complex and denote by $\bar{\mathcal{H}}$ its set of hyperplanes. Suppose that $\bar{\mathcal{H}}$ is a disjoint union of subsets $\bar{\mathcal{H}}_1$ and $\bar{\mathcal{H}}_2$, that is, $\bar{\mathcal{H}} = \bar{\mathcal{H}}_1 \sqcup \bar{\mathcal{H}}_2$. Suppose further that every pair $h_1 \in \bar{\mathcal{H}}_1$ and $h_2 \in \bar{\mathcal{H}}_2$ has nonempty intersection, i.e. $h_1 \cap h_2 \neq \emptyset$. Then $X \cong X_1 \times X_2$ with $X_i = X°(\mathcal{H}_i)$ for $i = 1, 2$.*

Proof Roller duality, i.e., Theorem 5.27, implies that $X°(\mathcal{H}(X)) \cong X$. It is therefore enough to prove that the half-space systems $\bar{\mathcal{H}}$ and $\mathcal{K} := \mathcal{H}(X_1 \times X_2)$ are equal.

A hyperplane $k \in \bar{\mathcal{K}}$ is defined by an edge e in $X_1 \times X_2$ and corresponds to the union of all midcubes $M \pitchfork [e]$.

Let M be such a midcube in a cube $C \subset X_1 \times X_2$. Then $C = C_1 \times C_2$ where C_i is a cube in X_i. See Fig. 5.18 for a picture. Hence there are edges e_i in X_i such that all edges in $[e] \cap C$ are of the form $e = e_1 \times \{v\}$ with v a vertex in X_2.

Fig. 5.18 Product
decomposition of a cube
complex

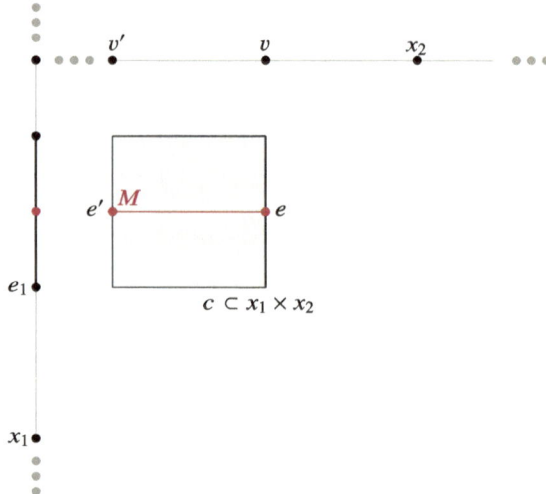

But, this implies that $H(e) \cong H(e_1) \times X_2$ and we may consider the map

$$\phi : \bar{\mathcal{K}} \longrightarrow \bar{\mathcal{H}} : H(e) \mapsto H(e_1).$$

Since every edge in $X_1 \times X_2$ is either of the form $e_1 \times \{v_2\}$ or $\{v_1\} \times e_2$ with v_i vertices in X_i and e_i edges in $X \setminus I$, one has that ϕ is surjective.

To prove that ϕ is also injective let e and f be two square equivalent edges is $X_1 \times X_2$. Then one of the two following cases occurs. Either $e = e_1 \times \{v_2\}$ and $f = e_1 \times \{v_2'\}$, that is, the edge in X_1 agrees, or $e = \{v_1\} \times e_2$ and $f = \{v_1'\} \times e_2$ and the edge in X_2 agrees. But, this implies that $\phi(e) = \phi(f)$ if and only if $e \sim f$ and the map ϕ is injective.

From what we have discussed the geometric realization of the elements in $\bar{\mathcal{K}}$ are of the form $H(e_i) \times X_j, i \neq j, i, j \in \{1, 2\}$. Recall the finite interval condition from Definition 5.10 item 2. For $\mathcal{K}$ this condition is inherited from the same property of $\mathcal{H}_1$ and $\mathcal{H}_2$.

The nesting condition, Definition 5.10 item 3, also follows from the nesting condition for the $\mathcal{H}_i$, $i = 1, 2$, and the fact that every $h_1 \in \mathcal{H}_1$ is transversal to all $h_2 \in \mathcal{H}_2$.   □

The converse of Theorem 5.31 is also true. See Exercise 5.41. This motivates the next definition.

Definition 5.32 (Irreducible Cube Complex) A CAT(0) cube complex X that can not be written as a product of two non-trivial CAT(0) cube complexes is called *irreducible*.

As one would expect, every CAT(0) cube complex can be decomposed into irreducible factors.

Theorem 5.33 (Decomposition Theorem) *Let X be a finite dimensional* CAT(0) *cube complex. Then, up to permutations of factors, the space X has a unique decomposition as a product of irreducible factors.*

Proof Let $X = \prod_{i=1}^{n} X_i$ be a maximal decomposition of X into n irreducible factors where n is at most the dimension of X. Let $X = \prod_{j=1}^{m} Y_j$ be a different maximal decomposition into irreducible components. Exercise 5.41 implies that

$$\bar{\mathcal{H}} = \bigcup_{i=1}^{n} \bar{\mathcal{H}}_i = \bigcup_{j=1}^{m} \bar{\mathcal{K}}_j$$

where the sets $\bar{\mathcal{H}}_i$ are pairwise disjoint and transversal as are the sets $\bar{\mathcal{K}}_j$. By assumption the sets $\mathcal{H}_i$ and $c\bar{\mathcal{K}}_j$ can not be decomposed.

But then, for every i there must be a j such that $\bar{\mathcal{H}}_i \subset \bar{\mathcal{K}}_j$ and for every k there must be an l such that $\bar{\mathcal{K}}_k \subset \bar{\mathcal{H}}_l$. Observe that the union over all the $\bar{\mathcal{K}}_j$ equals the union over all $\bar{\mathcal{H}}_i$. We hence obtain that $m = n$ and that there must be a permutation σ of $\{1, \ldots, n\}$ such that $\bar{\mathcal{H}}_{\sigma(i)} = \bar{\mathcal{K}}_i$. □

There are many ways of making sense of a boundary for CAT(0) spaces. Roller duality adds to this and provides a combinatorial way of defining a boundary for CAT(0) cube complexes called the Roller boundary.

Definition 5.34 (The Roller Boundary) The *Roller boundary* of a CAT(0) cube complex is the set

$$\partial_R Y := X(\mathcal{H}(Y)) \setminus X^\circ(\mathcal{H}(Y)).$$

Example 5.35 The Roller boundary of the standard cubing of the Euclidean plane is the union of four vertices with four infinite, simplicial lines as shown in Fig. 5.19.

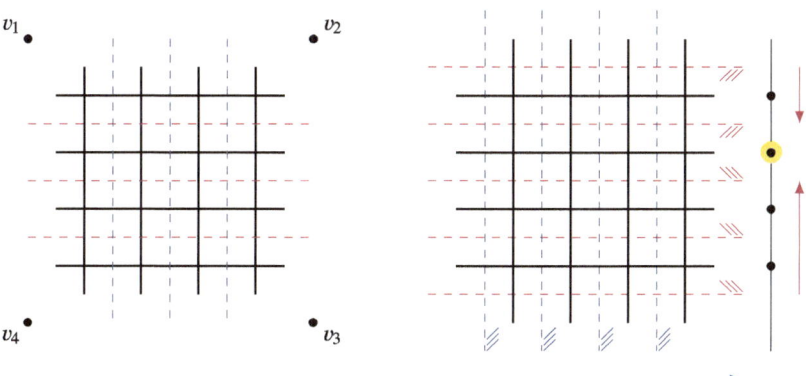

Fig. 5.19 Sageev's cubulation method also constructs additional points (left) and cube complexes (right) at infinity

The four vertices are labeled $v_1, \ldots, v_4$ on the left-hand side of the figure. One of the four lines in shown on the right-hand side of the figure. Compare Example 5.23, where we discuss how these vertices at infinity arise.

We close this section with a couple of remarks on more recent work concerning this boundary of a CAT(0) cube complex.

Beyrer et al. [BFIM21] introduce a cross ratio on the Roller boundary and show that CAT(0) cube complexes are in fact determined by these cross ratios—much like spherical buildings at infinity of affine buildings together with a valuation determine the affine building. Recently the connection between the Roller boundary and the Poisson boundary, studied by Nevo-Sageev [NS13], has been studied by Levovitz [Lev21]. A different notion, the simplicial boundary, has been studied by Hagen [Hag13, Hag18].

One can define a topology on the Roller boundary $\partial_R Y$ as follows. For every half-space h let U_h be the set of all vertices v in the vertex set of $(X(\mathcal{H}))$ for which $v(\bar{h}) = h$. Then take $\{U_S | S \subset \mathcal{H}\}$ as a sub-basis of a topology. The resulting topology is the same as the metric topology one obtains from the path-metric d_1 on the 1-skeleton of $X(\mathcal{H})$. For details see Section 2.3 in [NS13] or [Rol98].

Other notions of boundaries may also apply to CAT(0) cube complexes. Examples are the Tits boundary ∂X of a CAT(0) space X, which is, as a set, the collection of all equivalence classes of geodesic rays up to bounded distance. The simplicial boundary of a CAT(0) cube complex is similar to the Tits boundary, but geodesics are taken in the 1-skeleton here. See [Hag13].

5.4   Exercises

Exercise 5.36   Use Gromov's link condition to show that hyperplanes in a CAT(0) cube complex are locally CAT(0).

Exercise 5.37   Consider a topological surface S of genus two obtained as a quotient of a regular octagon A. We equip the octagon with a cube structure as follows: The midpoint of A forms a vertex v. Subdivide the 1-dimensional sides of A barycentrically and add an edge $\{u, v\}$ for each of the new barycenters w. Let then X be the universal cover of S by the cube structure induced by the cubulation of A we have just described. Show that X is a cube complex and that the fundamental group $G = \pi_1(S)$ of S acts cellularly and by isometries on X.

Exercise 5.38   Let X be the cube complex of Exercise 5.37. Answer the following questions.

1. What are the hyperplanes in X and how do the orbit quotients of hyperplanes look like? Are the quotients locally finite?
2. Describe the induced G-action on the orbit quotient and decide whether it is properly discontinuous or not.

3. The group G acts on the product of the orbit quotients. Describe the action and decide whether it is properly discontinuous and cocompact.

We need a new definition for some of the exercises.

Definition 5.39 (Arrow Diagram) Let $(P, \preceq)$ be a (finite) partially ordered set. The *arrow diagram* of $(P, \preceq)$ is a directed graph with vertex set P. Two vertices $x, y \in P$ are connected by a directed edge $y \rightarrow x$ if $x \preceq y$ and if $x \preceq z \preceq y$ implies that $z \in \{x, y\}$.

Note that the arrow diagram is a directed version of the Hasse diagram of a partially ordered set.

Exercise 5.40 Let $\mathcal{H}$ be the set containing the first twelve letters of the alphabet. The following arrow diagram defines a partial order on the set $\mathcal{H}$.

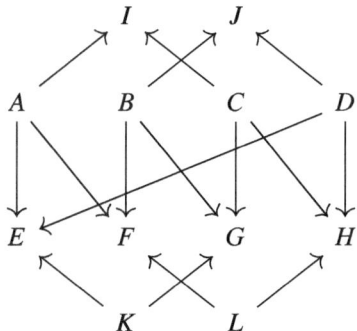

Decide whether or not $(\mathcal{H}, \preceq)$ can be extended to a half-space system, that is, whether there exists an order reversing map $\star$ such that the triple $(\mathcal{H}, \preceq, \star)$ satisfies all the properties of a half-space system.

Exercise 5.41 Let $X = X_1 \times X_2$ be the product of two CAT(0) cube complexes and let $\mathcal{H}_i = \mathcal{H}(X_i)$ be the half-space system of X_i for $i = 1, 2$. Show that the half-space system of X is given by $\mathcal{H}(X) = \mathcal{H}_1 \bigsqcup \mathcal{H}_2$ where for each pair of half-spaces $h_1 \in \mathcal{H}_1$ and $h_2 \in \mathcal{H}_2$ one has that $h_1 \pitchfork h_2$.

Exercise 5.42 A half-space system $(\mathcal{H}, \preceq, \star)$ is given by the set

$$\mathcal{H} = \{A, B, C, D, E, A^\star, B^\star, C^\star, D^\star, E^\star\}$$

and the relations $B \preceq A$, $B \preceq C$, $E \preceq D \preceq C$ and $A^\star \preceq E$. Draw the associated arrow diagram and the CAT(0) cube complex $X(\mathcal{H})$ it describes.

Exercise 5.43 Describe the half-space system of the CAT(0) cube complex pictured in Fig. 5.20 below using an arrow diagram.

Fig. 5.20 A cube complex

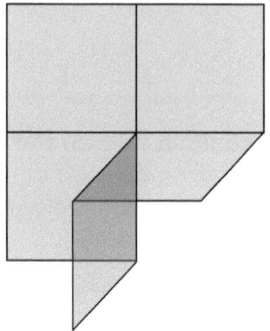

Exercise 5.44 Let $(\mathcal{H}, \preceq, \star)$ be a half-space system of finite width. Show that every two vertices in the associated cube complex $X^{\circ}(\mathcal{H})$ are connected by an edge-path in the 1-skeleton of $X^{\circ}(\mathcal{H})$.

Exercise 5.45 Prove Lemma 5.25.

Exercise 5.46 Let X be a finite dimensional CAT(0) cube complex equipped with the standard metric d. Show that

1. there exists a constant $R > 0$ such that every set of at least R hyperplanes in X contains at least two hyperplanes that have empty intersection.
2. X is quasi-isometric to its 1-skeleton $X^{(1)}$ equipped with the d_1-metric.

Exercise 5.47 The Coxeter complex Σ of the Coxeter group

$$W = \langle r, s, t \mid r^2, s^2, t^2, (rs)^3, (st)^3, (rt)^3 \rangle$$

of type $\tilde{A}_2$ is the simplicial complex underlying the tessellation of the Euclidean plane by equilateral triangles of side length one, a piece of which is shown in Fig. 5.21. Solve the following exercises.

Fig. 5.21 A piece of the Coxeter complex of type $\tilde{A}_2$

1. What are the hyperplanes and half-spaces in this complex Σ describing a half-space system for the underlying Coxeter group?
2. Describe the CAT(0) cube complex $X(\mathcal{H})$ for the half-space system found in item 1. How can one "see" the cubes in the triangulation of the plane?
3. $X(\mathcal{H})$ is isomorphic to a well known (3-dimensional) cube complex. Which one?
4. Describe the action of W on Σ and $X(\mathcal{H})$.

This construction of Exercise 5.47 is carried out in detail in [VSvRK21].

Chapter 6
Cubulating Coxeter Groups

Coxeter groups are abstract versions of reflection groups. They are widely studied in many areas of mathematics and play a central role in geometric group theory. In this chapter we quickly introduce Coxeter groups and some of their major properties. Moreover, we construct half-space systems for Coxeter groups. This allows us to apply Sageev's cubulation method (see Chap. 5) to this class of groups and obtain cube complexes on which the Coxeter groups act.

The natural candidates for the hyperplanes and half-spaces are given by the reflection hyperplanes of the Coxeter group. We define a family of half-spaces as subsets of the Cayley graph with respect to the Coxeter generators and show that it does indeed satisfy the requirements of an abstract half-space system as introduced in Sect. 5.2. The construction of the half-space system provided in this chapter is based on work of Niblo and Reeves [NR03]. We give a partially new proof of their main result which is more straightforward and shorter.

The first section contains the definition of Coxeter groups and associated notions such as Coxeter matrices, diagrams and Complexes. Several examples are provided. In the second section we prove that Coxeter groups satisfy the deletion condition. Finally in Sects. 6.3 and 6.4, we introduce walls and the associated half-space systems.

6.1 Coxeter Groups: Definition and Examples

Coxeter groups are defined by means of explicit presentations using generators and relations. They are abstract versions of reflection groups in spheres, Euclidean and hyperbolic space. In this section Coxeter groups are introduced and some first examples are provided.

Those Coxeter groups which admit realizations as (finite or infinite) Euclidean reflection groups are fully classified and come as a small list of infinite families

© The Author(s), under exclusive license to Springer Nature Switzerland AG 2023   109
P. Schwer, *CAT(0) Cube Complexes*, Lecture Notes in Mathematics 2324,
https://doi.org/10.1007/978-3-031-43622-2_6

together with a finite list of exceptional cases. A proof of the classification can, for example, be found in Humphreys' book [Hum72]. Other classic books on Coxeter groups include [Dav98, BB05, Bou02] or the first half of [Bro89]. For a quick introduction and many pictures see also the first section in [Sch22].

Definition 6.1 (Coxeter Matrices) A *Coxeter matrix* $M = (m_{ij})_{i,j}$ is an $n \times n$ symmetric matrix with entries in $\mathbb{N} \cup \{\infty\}$ such that $m_{ij} = 1$ if $i = j$ and such that $2 \leq m_{ij} \leq \infty$ for all other i, j.

Given a Coxeter matrix we can define a Coxeter group.

Definition 6.2 (Coxeter Systems and Groups) Let $M = (m_{ij})_{i,j}$ be a Coxeter matrix of size $n \times n$. Define a group W by means of the following presentation: Let $S = \{s_1, \ldots, s_n\}$ be a set of abstract generators and put

$$W = \langle s_1, \ldots, s_n \mid (s_i s_j)^{m_{ij}} = 1, i, j \in \{1, 2, \ldots, n\}\rangle,$$

where $m_{i,j} = \infty$ means there is no relation between the elements s_i and s_j. The pair (W, S) is called a *Coxeter system* of type M.

A group G is a *Coxeter group* if there exists a Coxeter matrix M with Coxeter system (W, S) and G is isomorphic to W as a group. We call $n = |S|$ the *rank* of W.

Suppose s_i and s_j are two reflections along two intersecting hyperplanes in Euclidean space. Then the order $m_{i,j}$ of the product $s_i s_j$ encodes the angle α in which the associated reflection hyperplanes meet: $\alpha = \frac{\pi}{m_{ij}}$. The definition of a Coxeter group generalizes this geometric interpretation to abstract groups.

When writing down the presentation of a Coxeter group we may leave out trivial relations of the form $(s_i s_j)^\infty$ and only record one of the two identical relations $(s_i s_j)^{m_{i,j}}$ and $(s_j s_i)^{m_{j,i}}$.

The Coxeter matrix and also the presentation of a Coxeter group W with respect to a fixed Coxeter generating set S can be encoded by a labeled graph.

Definition 6.3 (Coxeter Diagrams[1]) Let (W, S) be a Coxeter system with Coxeter matrix $M = (m_{ij})_{i,j}$. The *Coxeter diagram* of (W, S) is the graph Γ defined as follows: There exists a vertex i in Γ for each element s_i in S. Two vertices i and j are connected by an edge if and only if $m_{ij} \geq 3$. An edge between i and j is labeled with m_{ij} in case $m_{ij} \geq 4$.

Remark 6.4 As a set the group W consists of all finite words in the alphabet S where the empty word corresponds to the identity. Two given words x and y define the same element in W if and only if they can be transformed into one another by using a finite sequence of the following replacements: delete a pair $s_i s_i$ for $s_i \in S$, introduce a pair $s_i s_i$ at some position in the word for $s_i \in S$ or replace a string

---

[1] Coxeter diagrams are also known as Dynkin- or Coxeter-Dynkin diagrams.

Fig. 6.1 Coxeter diagram for the symmetric groups on n letters

$s_i s_j s_i \ldots$, where the expression has m_{ij}-many letters, by the string $s_j s_i s_j \ldots$ with the same amount of letters.

Here is a first example of a class of Coxeter matrices, groups and diagrams.

Example 6.5 (Coxeter Groups of Type A_n) Let M_n be the $n \times n$- matrix where $m_{i,i} = 1$, $m_{i,i+1} = m_{i+1,i} = 3$ for all $i = 1, \ldots, n-1$, and $m_{i,j} = 2$ otherwise. For $n = 2$ and 3 we obtain

$$M_2 = \begin{pmatrix} 1 & 3 \\ 3 & 1 \end{pmatrix} \text{ and } M_3 = \begin{pmatrix} 1 & 3 & 2 \\ 3 & 1 & 3 \\ 2 & 3 & 1 \end{pmatrix}$$

The Coxeter system (W_n, S_n) with respect to M_n yields a group W_n generated by the elements in S_n. The group W_n has the following presentation:

$$W_n = \langle S_n \mid s_i^2 \; \forall i, (s_i s_{i+1})^3 \; \forall \; 1 \le i \le n-1 \text{ and } (s_i s_j)^2 \; \forall \; |i-j| > 1 \rangle,$$

where we leave out the trivial relations $(s_i s_j)^\infty$ for $|i-j| > 1$.

The group W_n is isomorphic to the symmetric group $\mathrm{Sym}(n+1)$ on $n+1$ letters. The generators $s_i \in S_n$ correspond to the transposition $t_{i,i+1}$ in $\mathrm{Sym}(n+1)$ swapping i with $i + 1$. The Coxeter diagram of this system (W_n, S_n) is a graph Γ with n consecutive vertices in a row connected by unlabeled edges as shown in Fig. 6.1.

Geometrically the symmetric group $\mathrm{Sym}(n + 1)$ corresponds to the symmetry group of the regular simplex on $n+1$ vertices in $\mathbb{R}^n$. So each element of $\mathrm{Sym}(3)$ can be seen as a symmetry of an equilateral triangle in the Euclidean plane permuting the three vertices. An element of order two in this group corresponds to a reflection through a line that contains a vertex and the midpoint of the opposite side.

Here are some more examples of Coxeter groups.

Example 6.6 Symmetry groups of regular polytopes are Coxeter groups.

1. We have already seen one class of this kind, namely the symmetric groups $\mathrm{Sym}(n + 1)$, which are the symmetry groups of the regular simplices on $n + 1$ vertices. We refer to this class as of *type A_n*, where the n is the same as n in $\mathrm{Sym}(n + 1)$. Their presentations are given in Example 6.5.
2. The class of Coxeter groups *of type BC_n* for $n \ge 3$ contains all symmetry groups of n-cubes. Their Coxeter presentations are given by

$$W = \langle s_1, \ldots, s_n \mid s_i^2, \forall i, (s_j s_{j+1})^3, \text{ for all } 1 \le j \le n-2 \text{ and } (s_{n-1} s_n)^4 \rangle.$$

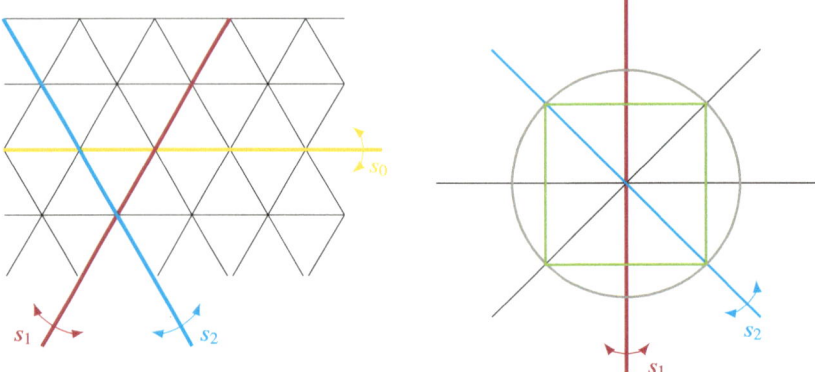

Fig. 6.2 Coxeter groups of type $\tilde{A}_2$ and $I_2(4)$ as real reflection groups in $\mathbb{R}^2$

3. The symmetry group of the icosahedron is given by

$$W = \langle s_1, s_2, s_3 \mid s_i^2, i = 1, 2, 3, (s_1 s_2)^5, (s_2 s_3)^3 \rangle$$

   and is referred to as *of type H_3*.
4. The finite dihedral groups, that is, the Coxeter groups of *type $I_2(n)$* with $n \geq 2$, are given by the following presentation:

$$W = \langle s_1, s_2 \mid s_1^2, s_2^2, (s_1 s_2)^{m_{1,2}} \rangle,$$

   with $m_{1,2} = n$. These groups may be visualized as finite reflection groups in the Euclidean plane generated by two reflections at an angle of $\frac{\pi}{n}$. See Fig. 6.2 for an illustration of the Coxeter group of type $I_2(4)$ as a reflection group of both $\mathbb{S}^1$ and $\mathbb{R}^2$.

For details on items 2 and 3 we refer to the book by Borovik and Borovik [BB10] on the classification of finite Coxeter groups. This book contains many beautiful examples of (finite) Coxeter groups.

Observe that all groups mentioned in this example are finite.

One infinite group that is similar to the finite dihedral groups discussed above is the infinite dihedral group D_∞. One may obtain this group generated by a pair of reflections along parallel lines in the Euclidean plane. Indeed this group is a Coxeter group with presentation $\langle s_1, s_2 \mid s_1^2, s_2^2 \rangle$. See the next example for more instances of infinite Coxeter groups.

Example 6.7

1. The infinite dihedral group is the smallest (in rank) example of an affine Coxeter group of *type $\tilde{A}_n$*. Here $n = 1$. The groups of type $\tilde{A}_n$ for $n \geq 2$ have the following presentation and their Coxeter diagram is a cycle consisting of $n + 1$

vertices and $n + 1$ unlabeled edges.

$$W = \langle s_0, s_1, \ldots, s_n \mid s_i^2, (s_i s_{i+1})^3, \forall i = 0, \ldots, n \text{ and } (s_i s_j)^2 \text{ for } |i - j| > 1 \rangle,$$

where we take indices modulo $n + 1$. The infinite Coxeter group of type $\tilde{A}_2$ is generated by three reflections pairwise intersecting at an angle of $\frac{\pi}{3}$ as illustrated on the left-hand side of Fig. 6.2. This group is the symmetry group of the equilateral triangle tiling of the plane.

2. Any triple of integers k, l, m with $\frac{1}{k} + \frac{1}{l} + \frac{1}{m} < 1$ gives rise to a reflection subgroup W of the automorphisms group of the hyperbolic plane. The three reflections generating the group W are along the three sides of a hyperbolic triangle with interior angles $\frac{\pi}{x}$ with $x \in \{k, l, m\}$ and has presentation

$$W = \langle s_1, s_2, s_3 \mid s_i^2, i = 1, 2, 3 \text{ and } (s_1 s_2)^k, (s_1 s_2)^l, (s_1 s_2)^m \rangle.$$

Another instance of such a group is discussed in Example 6.35 in Sect. 6.4.

We have so far only seen a few examples of Coxeter groups. In fact one can classify those acting cocompactly on either a sphere or some Euclidean space. See, for example, the book of Bjoerner and Brenti [BB05, p. 296] or Humphreys' book [Hum72]. In the following we occasionally use some of the terminology introduced in this classification such as names of types and diagrams as done above. One thing that is worth mentioning though is that all the affine Coxeter groups have a spherical counterpart that appears as a special subgroup and whose diagram is obtained from the affine diagram by deleting a vertex in such a way that the diagram remains connected.

Remark 6.8 (The Isomorphism Problem for Coxeter Groups) Note that there exist groups G, which are Coxeter groups in many different ways. That is, there may exist two (or more) Coxeter matrices and hence different Coxeter systems (W_1, S_1) and (W_2, S_2) with generating sets S_1 and S_2 such that G is isomorphic to both W_1 and W_2. The two groups may be of the same, but potentially also of non-equal rank.

A classical example of such a pair of Coxeter systems for the same group is the dihedral group of order 6, denoted D_6, and the product $S_2 \times S_3$ of two symmetric groups of rank 2 and 3. These two Coxeter groups are isomorphic as (abstract) groups.

For more details on the isomorphism problem and a summary of the current state of the art see [RS22].

The next goal is to construct for every Coxeter group a simplicial complex on which the group acts. This complex is defined from the so called special cosets in Coxeter groups. It is the order complex of the poset of special cosets in a Coxeter group W ordered by reverse inclusion. Here is the formal definition:

Definition 6.9 (Coxeter Complex) Let (W, S) be a Coxeter system. A *special subgroup* in W is a group $W_T < W$ generated by a proper subset T of S. Here

$W_\emptyset = \{\mathbb{1}\}$ the trivial group. A *special coset* wW_T in W is a left-coset of a special subgroup.

The *Coxeter complex* $\Sigma = \Sigma(W, S)$ of (W, S) is the simplicial complex defined as follows: The vertices of the complex are the special cosets wW_T with $|T| = |S| - 1$. Any two of them are connected whenever they contain a common special coset of the form $wW_{T'}$ with $|T'| = |S| - 2$. We identify the edge with this type of coset. Higher dimensional simplices are spanned by cliques in the resulting graph that correspond one-to-one to special cosets in W. A *face* of a simplex A in $\Sigma(W, S)$ is a proper sub-simplex of A.

In other words $\Sigma(W, S)$ is, as an abstract simplicial complex, the poset of standard cosets in W, ordered by reverse inclusion. Thus, a coset $wW_T \geq uW_{T'}$ in $\Sigma(W, S)$ if and only if $wW_T \subset uW_{T'}$ as subsets of W.

Remark 6.10 Maximal simplices in the Coxeter complex are all of the same dimension, namely $|S| - 1$, and are in bijection with special cosets of the form $wW_\emptyset$ with $w \in W$. Every codimension one face of a simplex is contained in exactly two simplices.

Example 6.11 Let $W = \mathrm{Sym}(3)$ be the symmetric group of three elements generated by two elements s_1 and s_2 of order 2. Then W has Coxeter presentation

$$W = \langle s_1, s_2 \mid s_1^2, s_2^2, (s_1 s_2)^3 \rangle$$

and has special subgroups $W_\emptyset$, $W_{\langle s_1 \rangle}$ and $W_{\langle s_2 \rangle}$. The special cosets are these groups and all their left cosets. The resulting Coxeter complex is a hexagon. See Fig. 6.3 for an illustration of the resulting Coxeter complex. Edges are labeled by cosets of the trivial subgroup and are hence in bijection with the elements of the group. We denote by w_0 the longest element $w_0 = s_1 s_2 s_1 = s_2 s_1 s_2$ in W.

Fig. 6.3 The Coxeter complex of the symmetric group S_3

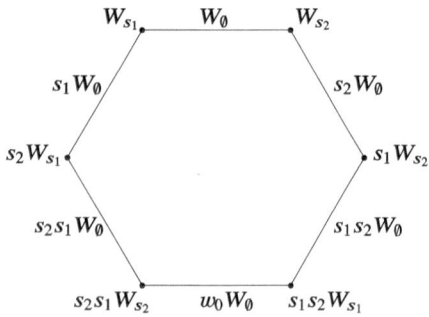

6.2   The Deletion Condition

One can show that a group generated by involutions is a Coxeter group whenever it satisfies the deletion condition. So this condition characterizes Coxeter groups and is crucial in many proofs. We will make use of it in proving the existence of a half-space system for Coxeter groups. In this section we introduce the deletion condition and prove that Coxeter groups have this property. Note that there are other, equivalent, characterizing conditions on Coxeter groups, such as the exchange condition. We refer the reader to e.g. [Bro89] to learn more about those and how to prove their equivalence.

Definition 6.12 (Deletion Condition) Let W be a group generated by a finite set S. Then the pair (W, S) satisfies the *deletion condition* if the following holds:

(D) If $w = s_1 \ldots s_m$ with $m > l_S(w)$ then there exists $i < j$ such that $w = s_1 \ldots \widehat{s_i} \ldots \widehat{s_j} \ldots s_m$.

Properties equivalent to the deletion condition are the *exchange condition*:

(E) Let $w \in W$ and $s \in S$ and suppose that w has the reduced presentation $w = s_1 \ldots s_d$. Then either $l_S(sw) = d + 1$ or there exists an i such that $w = ss_1 \ldots \widehat{s_i} \ldots s_d$.

and the *folding condition*:

(F) Let $w \in W$, $s, t \in S$ with $l_S(sw) = l_S(w) + 1$ and $l_S(wt) = l_S(w) + 1$ then either $l_S(swt) = l_S(w) + 2$ or $swt = w$.

Proofs that these conditions are equivalent and satisfied by all Coxeter groups are provided in [AB08] or [Dav98, Thm 3.3.4 p.38]. We prove that Coxeter groups satisfy the deletion condition in Theorem 6.20.

The next lemma is a consequence of the deletion condition (D). It contains a nice property of the standard length function of a Coxeter system. See also Exercise 6.44.

Lemma 6.13 ([Ron89, 2.4]) *Let (W, S) be a Coxeter group and u, v adjacent vertices in its Cayley graph Γ_W. Then $d_S(x, u) = d_S(x, v) \pm 1$ for all $x \in W$.*

Proof A minimal presentation of $x^{-1}u = s_1 \ldots s_d$ corresponds to a geodesic path γ in Γ_W from x to u consisting of edges that are labeled with s_1, s_2, etc. And the vertices $x, xs_1, xs_1s_2, \ldots, u = xs_1 \cdots s_d$ appear along γ in this order.

Elongate γ by one edge with label s so that it goes to $v = us$. This corresponds to the product $x^{-1}us$. Now either $l(x^{-1}v) = l(x^{-1}u) + 1$ or $l(x^{-1}v) < l(x^{-1}u)$. From (D) we can conclude that there exist two letters in the presentation, which may be deleted, and $l(x^{-1}v) = l(x^{-1}u) - 1$, which finishes the proof.   □

Following Davis [Dav98] we show that Coxeter groups satisfy (D).

Definition 6.14 (Reflections) A *reflection* r in a Coxeter system (W, S) is a conjugate wsw^{-1} of a generator $s \in S$ for some arbitrary $w \in W$. The set

$R = R(W, S)$ of all reflections is the collection of all conjugates of generators $s \in S$ by elements $w \in W$.

It is clear from the definition that the group W acts by conjugation on the set R of its reflections. Moreover, all reflections are elements of order two. However, the converse is not true. In general, the set of reflections depends on the generating set S and can be strictly smaller than the set containing all involutions.

Lemma 6.15 *Let Γ be the Cayley graph of W with respect to S. Then*

1. *for every edge $\{u, v\}$ in Γ there exists a reflection $r \in R$ such that $ru = v$ (and thus also $rv = u$), and*
2. *for every $r \in R$ there exists an edge $e = \{u, v\}$ in Γ such that $ru = v$ (and thus also $rv = u$).*

Proof To prove the first item we may assume that $u = vs$ for some $s \in S$. Put $r := vsv^{-1}$. Then $ru = v$ and $rv = vs = u$. To see the second item fix some edge $\{u, v\}$ with $u = vs$ and choose $r = usu^{-1}$. $\qquad\square$

Associated with a given word in a Coxeter group is a list of reflections.

Definition 6.16 (Reflections of a Word) Let $\bar{w}$ be a word in S, that is, $\bar{w} = s_1 \cdot \ldots \cdot s_k$ with $s_i \in S$ for all $i = 1, \ldots, k$. Put $w_0 = \mathbb{1}$ and $w_i = s_1 \cdot \ldots \cdot s_i$ for all $1 \leq i \leq k$. The *reflections of w* are the elements $r_1, \ldots, r_k$ defined by $r_i := w_{i-1} s_i w_{i-1}^{-1}$ with $1 \leq i \leq k$.

Note that the list of reflections of a given $\bar{w}$ might have duplicates. By construction the product of the first i generators in the word equals the product of the first i reflections, that is, $s_1 \cdot \ldots \cdot s_i = w_i = r_i \cdot \ldots \cdot r_1$ for all $1 \leq i \leq k$.

Lemma 6.17 *Let $\bar{w} = s_1 \cdot \ldots \cdot s_k$ be a word in S representing an element $w \in W$ and let $r_1, \ldots, r_k$ be its reflections. Suppose $r_i = r_j$ for some $i \neq j$. Then the edge-paths associated with the words $\bar{w}$ and $\bar{u} := s_1 \cdot \ldots \cdot \widehat{s_i} \cdot \ldots \cdot \widehat{s_j} \cdot \ldots \cdot s_k$ in Γ have the same endpoints and $u = w$ in W.*

Proof The sequence of elements w_i with $1 \leq i \leq k$ defined in Definition 6.16 are the vertices of an edge-path γ in Γ from $\mathbb{1}$ to w. Suppose that $i < j$. Since $r_i = r_j$ the reflection r_i switches the two vertices of the edge $e = \{w_{i-1}, w_i\}$ and also the vertices of the edge $e' = \{w_{j-1}, w_j\}$. Since every reflection preserves adjacency of vertices in Γ the simplicial map $r_i : \Gamma \to \Gamma$ maps the piece τ of γ connecting w_i with w_{j-1} to an edge-path $\tau' := r(\tau)$ connecting w_{i-1} with w_j. Compare Fig. 6.4.

Replace γ by the following path: take the sub-path of γ connecting $\mathbb{1}$ with w_{i-1}, connect it to τ' and then to the sub-path of γ that connects w_j with w_k. This path corresponds by construction to the word $\bar{u}$ and is two edges shorter than γ.

It remains to prove that u equals w. Proceed as follows: The fact that $r_i = r_j$ is equivalent to $s_1 \cdots s_{i-1} s_i s_{i-1} \cdots s_1 = s_1 \cdots s_{j-1} s_j s_{j-1} \cdots s_1$. But then, $s_i \cdots s_j = s_{i+1} \cdots s_{j+1}$ and we may replace the sub-word $s_i \cdots s_j$ in $\bar{w}$ by the word $s_{i+1} \cdots s_{j+1}$ without changing the corresponding element in W. This implies the assertion. $\qquad\square$

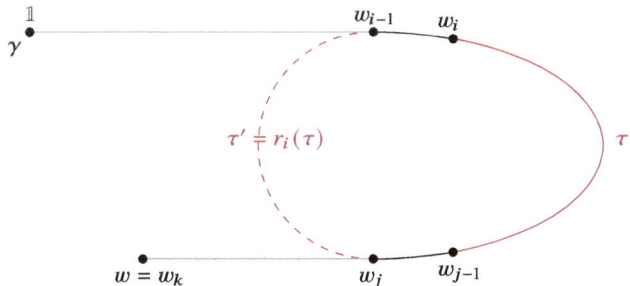

Fig. 6.4 Shortening an edge-path, which passes through the same reflection twice. A step in the proof of 6.17

We obtain the next statement as an immediate consequence.

Corollary 6.18 *Let $\bar{w}$ be a word representing a minimal edge-path in the Cayley graph $\Gamma = \Gamma(W, S)$. Then the reflections of w are pairwise different, that is, the family of all reflections of w is a set (and has no duplicates).*

In fact, the converse of this property is also true.

Lemma 6.19 *Let $\bar{w} = s_1 \cdots s_k$ be a word in S representing an element $w \in W$. Suppose that all reflections $r_1, \ldots, r_k$ of w are pairwise distinct. Then $\bar{w}$ is reduced and represents a minimal edge-path in $\Gamma(W, S)$.*

We provide a proof of this fact at the end of Sect. 6.3 once we have introduced some additional material. In Definition 6.21 we associate to a reflection r in a Coxeter group W a subset of the edges of the Cayley graph $\Gamma(W, S)$, called wall of r. It is not hard to see that the statement in Lemma 6.19 can be refined to the following fact: the reflections r_i in a (minimal) word are in natural bijection with the walls crossed by the associated (minimal) edge-path. See Definition 6.22 for the definition of a crossed wall.

Theorem 6.20 (Coxeter Groups Satisfy the Deletion Condition) *Every Coxeter group (W, S) satisfies property (D).*

Proof Let $\bar{w} = s_1 \cdots s_k$ be a word for $w \in W$. If $\bar{w}$ is not reduced then there exist, by Lemma 6.19, indices $1 \leq i \neq j \leq k$ such that the reflections r_i and r_j are equal. Lemma 6.17 then implies that $w = s_1 \cdots \widehat{s_i} \cdots \widehat{s_j} \cdots s_k$ and (D) follows. □

6.3 Walls in Coxeter Groups

The goal of this section is to apply the cubulation procedure introduced in Sect. 5.2, and in particular Theorem 5.15, to Coxeter groups. We therefore define walls and half-spaces in Coxeter groups.

Fig. 6.5 The wall M_t
consists of two edges $\{\mathbb{1}, t\}$
and $\{st, sts\}$. The two
associated roots (see
Definition 6.26) are the sets
consisting of the green,
respectively blue vertices

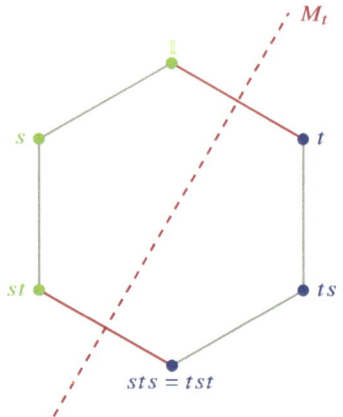

Intuitively, a wall is a reflection hyperplane or reflection hypersphere in a geometric realization of the associated Coxeter complex. Formally, walls and half-spaces are defined in terms of collections of edges in the Cayley graph.

Definition 6.21 (Walls of Reflections in Coxeter Groups) Let (W, S) be a Coxeter system and write R for the set of reflections in W. The *wall* M_r of a reflection $r \in R$ is the set of edges $\{u, v\}$ in the Cayley graph of W with $u = rv$.

Note that every edge $\{w, ws\}$ is contained in the wall M_r with $r = wsw^{-1}$. Moreover this wall is the unique wall containing this edge and there is a one-to-one correspondence between walls M_r and reflections wsw^{-1} in W.

Definition 6.22 (Crossing Walls) Let $\gamma = (c_0, c_1, \ldots, c_k)$ be an edge-path in Γ_W. We say that γ *crosses* the wall M_r if there exists $1 \leq i \leq k$ such that the reflection r maps c_{i-1} to c_i and vice versa. Let $n(\gamma, r)$ denote the number of times γ crosses the wall M_r. We say that a wall M_r *separates* two vertices c and c' if there exists a minimal edge-path from c to c' that crosses M_r.

Example 6.23 Consider Sym(3), the symmetric group on $\{1, 2, 3\}$. This group is generated by the two transpositions $t_{1,2}$ and $t_{2,3}$ swapping 1 and 2 or 2 and 3, respectively. The group Sym(3) is abstractly isomorphic to the Coxeter group W given by the presentation $W = \langle s, t \mid s^2, t^2, (ts)^3 \rangle$. The isomorphism maps $t_{1,2}$ to s and $t_{2,3}$ to t,

A picture of the Cayley graph of W with respect to the generating set $S = \{s, t\}$ is shown in Fig. 6.5. There are three reflections in W: the two generators s, t and their conjugate $sts = tst$. Each reflection defines a wall consisting of two edges: $M_s = \{\{\mathbb{1}, s\}, \{ts, tst\}\}$, $M_t = \{\{\mathbb{1}, t\}, \{st, sts\}\}$ and $M_r = \{\{s, st\}, \{t, ts\}\}$. The wall M_t separates s from t. See also Example 6.27.

Lemma 6.24 *Let $\bar{w}$ be a word in the generators S of a Coxeter group W. The set of walls crossed by the edge-path defined by $\bar{w}$ is the collection of walls M_r, where r is a reflection of w as in Definition 6.16.*

Proof This is a direct consequence of item 1 of Lemma 6.15.   □

We prove, in the second assertion of the next lemma, that any edge-path connecting a pair of vertices x and y either always crosses a wall M_r an even number of times or an odd number of times but the parity is independent of the path. So, whether M_r does separate x and y or not can be decided by the parity of the number of crossings.

Lemma 6.25 *Let (W, S) be a Coxeter system and $\Gamma = \Gamma(W, S)$ its Cayley graph. Then*

1. *every minimal edge-path in Γ crosses each wall at most once, i.e.*

$$n(\gamma, r) \leq 1 \text{ for all reflections } r, \text{ and all minimal edge-paths } \gamma. \text{ And}$$

2. *for any pair of vertices x, y in Γ and any two edge-paths $\gamma, \gamma' : x \rightsquigarrow y$ one has*

$$n(\gamma, r) = n(\gamma', r) \mod 2.$$

Proof The first assertion follows from item 1 of Lemma 6.15. To see this argue as follows: let $\gamma = (c_0, c_1, \ldots, c_k)$ be a minimal edge-path in Γ and suppose there exists M_r, which is crossed by γ twice, say at positions i and j. Then the minimal path $(c_i, c_{i+1}, \ldots, c_{j-1})$ is mapped onto an edge-path of the same length by r, but now connecting $r(c_i) = c_{i-1}$ and $r(c_{j-1}) = c_j$. But then, we may shorten γ to the path $(c_1, \ldots, c_{i-1}, r(c_{i+1}), \ldots, r(c_{j-2}), c_j, c_{j+1}, \ldots, c_k$, which contradicts minimality of γ.

By transitivity of the left-action of G on its Cayley graphs it is enough to prove item 2 for the identity and any other element x in the group. Fix x in W and choose a (not necessarily reduced) word $s_1 \cdots s_d$ for x. Denote by $\gamma : \mathbb{1} \rightsquigarrow x$ the associated edge-path in Γ. Further let M_r be a wall in Γ.

Any two words for x in W may be transformed into one another by repeated application of the following moves:

1. Delete double appearance ss of a generator $s \in S$.
2. Insert double appearance ss of a generator $s \in S$.
3. Replace a string $s_i s_j s_i \cdots$ of m_{ij} letters with the string $s_j s_i s_j \cdots$ of m_{ij} letters.

We need to show that the application of any of these moves does not change the parity of $n(\gamma, r)$. Move 1 shortens the edge-path γ. Hence, if M_r is not crossed by γ it is also not crossed by the shortened path after deleting the repeated edge. This is also the case if the deleted edges were not contained in M_r. In case M_r was crossed by γ and the repeated edge lies in M_r, then the value of $n(\gamma, r)$ decreases by 2. Move 2 for the same reason also leaves the value of $n(\gamma, r)$ either unchanged or increases this value by $+2$. Move 3 does not change the value of $n(\gamma, r)$ as such a substitution yields a path, which crosses the same hyperplanes, but in a different order. One can illustrate this fact by considering an edge-path connecting $\mathbb{1}$ with the vertex labeled sts in Fig. 6.5. Such a path is e.g. given by the sequence of vertices $(\mathbb{1}, s, st, sts)$.

Move 3 replaces this path by the sequence $(\mathbb{1}, t, ts, tst)$ which crosses the same set of walls. Hence, the assertion.   □

Reflection hyperplanes in, for example, the Euclidean plane divide the plane into two separate half-spaces, which are interchanged by the reflection. This phenomenon is captured for abstract reflections in Coxeter groups by the next definition and its properties.

Definition 6.26 (Opposite Roots) Let (W, S) be a Coxeter system and r a reflection in W. Denote by M_r the wall associated with r in the Cayley graph Γ of (W, S). The *roots* associated with the reflection r are the two subsets $\pm\alpha$ of the vertices of Γ given by

$$\alpha = \{w \in W \mid n(\gamma, r) = 0 \mod 2 \text{ for every minimal edge-path } \gamma : \mathbb{1} \rightsquigarrow w\}$$

and

$$-\alpha = \{w \in W \mid n(\gamma, r) = 1 \mod 2 \text{ for every minimal edge-path } \gamma : \mathbb{1} \rightsquigarrow w\}.$$

We say that the roots α and $-\alpha$ are *opposite* in Γ.

By Lemma 6.25 roots are well defined as the parity of $n(\gamma, r)$ is independent of the choice of a connecting edge-path. Note also that for any choice of a reflection the vertices of Γ are precisely the union of α and $-\alpha$.

We provide next an example of walls and roots in a finite Coxeter group. For an example of an infinite group see Example 6.31.

Example 6.27 Let $W = \mathrm{Sym}(3)$ be the symmetric group on three elements. Its Cayley graph with respect to the standard Coxeter presentation is a hexagon shown in Fig. 6.5. There are three reflections in W: the two generators s, t and their conjugate $sts = tst$. Each reflection defines a wall consisting of two edges: $M_s = \{\{\mathbb{1}, s\}, \{ts, tst\}\}$, $M_t = \{\{\mathbb{1}, t\}, \{st, sts\}\}$ and $M_r = \{\{s, st\}, \{t, ts\}\}$. Figure 6.5 illustrates the wall M_t by connecting the two edges it contains with a dashed red line. The two roots defined by M_t are the two subsets of the vertices of the complex having the same color. That is, $\{\mathbb{1}, s, st\}$ and $\{t, ts, tst\}$.

The next lemma is crucial for the cubulation of Coxeter groups. Compare also [Ron89, Prop. 2.6].

The sets $H(u, v)$ defined in Lemma 6.28 allow us to construct half-space systems for Coxeter groups in the next subsection.

Lemma 6.28 *Let u, v be adjacent vertices in $\Gamma(W, S)$. Define the set $H(u, v)$ by putting*

$$H(u, v) := \{c \in W \mid d_S(c, u) < d_S(c, v)\}.$$

Every root α with $u \in \alpha$ and $v \in -\alpha$ satisfies the following three properties

1. $\alpha = H(u, v)$ and $-\alpha = H(v, u)$,
2. $H(u, v) = H(x, y)$ for any pair of adjacent vertices x, y with $x \in H(u, v)$ and $y \notin H(u, v)$,
3. $H(u, v) \cap H(v, u) = \emptyset$ and $W = H(u, v) \sqcup H(v, u)$.

Proof To show that $\alpha = H(u, v)$ we first prove that α is contained in $H(u, v)$. Let $c \neq u$ be a vertex in α. The definition of roots implies that a minimal edge-path from c to u crosses M_r an even number of times. By Lemma 6.25 item 1 it crosses at most once. Hence, a minimal edge-path does not cross the wall M_r at all. In particular such an edge-path can not contain v. But then, Lemma 6.13 implies that $d(c, v) = d(c, u) + 1 > d(c, u)$ and hence $c \in H(u, v)$.

To show that $H(u, v) \subset \alpha$ suppose now that $c \in H(u, v)$. We claim: there exists an edge-path connecting v and c, which goes via u. Such a path crosses M_r by construction and has crossing number at most one by Lemma 6.25 item 1. This implies $c \notin -\alpha$.

It remains to prove the claim that there exists an edge-path connecting v and c, which goes via u. Suppose $\gamma : c \rightsquigarrow u$ is minimal. Since u and v are adjacent we may deduce from Lemma 6.13 that

$$l(c^{-1}us) = l(c^{-1}v) = d(c, v) = d(c, u) + 1 = l(c^{-1}u).$$

The edge-path γ corresponds to a minimal presentation of $c^{-1}u$, say $s_1 \cdots s_d$. Then $s_1 \cdots s_d s$ is a minimal presentation of $c^{-1}v$. The elongation of γ by the edge $\{u, v\}$ is a minimal edge-path, which connects u with v and crosses M_r. And thus $\alpha = H(u, v)$.

One similarly verifies that $-\alpha = H(v, u)$. The second and third items are easy consequences of the definition of roots and the first item. □

The following property is used in the next section to show that the roots form a half-space system.

Lemma 6.29 *Let (W, S) be a Coxeter system and let $H_1 = H(\mathbb{1}, t)$ and $H_2 = H(u, v)$ be two roots as in Lemma 6.28. If $H_1 \subsetneq H_2$, then $t \in H_2$ and $u \notin H_1$.*

Proof By definition $t \notin H_1$ and $u \in H_2$. As $H_1 \subset H_2$ we may conclude that $\mathbb{1} \in H_2$. We argue by contradiction to prove that $t \in H_2$. Suppose $t \notin H_2$. Then it follows by Lemma 6.28 item 2 that $H_2 = H_1$. This contradicts the assumption that $H_1 \neq H_2$. Thus, $t \in H_2$.

To prove that $u \notin H_1$ we argue as follows. First observe that it can not happen that both u and v are contained in H_1 as this contradicts the fact that H_1 is a proper subset of $H_2 = H(u, v) \not\ni v$. So v can not be contained in H_1. Suppose now that $u \in H_1$ and $v \notin H_1$. Then again by item 2 of Lemma 6.28 we can conclude that $H_1 = H_2$ and we arrive at a contradiction. This implies that $u \notin H_1$. □

One may simultaneously view $H(u, v)$ as a subset of the group W and also as a subset of the vertices of the Cayley graph of W with respect to the generating set S.

We end this section with the still missing proof of Lemma 6.19. This completes the proof of the fact that Coxeter groups satisfy the deletion condition.

Proof of Lemma 6.19 Let $\bar{w} = s_1 \cdots s_k$ be a word in S representing an element $w \in W$. Suppose that all reflections $r_1, \ldots, r_k$ of w are pairwise distinct. We want to prove that $\bar{w}$ is reduced and represents a minimal edge-path in $\Gamma(W, S)$. It is clear from the definitions that the edge-path represented by a word is minimal if and only if the word is reduced. It is hence enough to prove that the path γ represented by $\bar{w}$ is minimal. Lemma 6.24 implies that the walls crossed by γ are exactly the walls associated with the reflections r_i of w. Hence Lemma 6.25 implies that an edge path is minimal if and only if it crosses every such wall at most once. For the walls M_{r_i}, with r_i a reflection of w, one has that $n(\gamma, r_i) \geq 1$ by construction, and we want to prove equality. Suppose for a contradiction that there exists a reflection $r \in \{r_1, \ldots, r_k\}$ such that $n(\gamma, r) > 1$. But then the wall M_r is crossed at least twice, and thus there exist two distinct indices $i \neq j$ such that $r_i = r_j$. But this contradicts the fact that all r_i are distinct. Hence the assertion follows.   □

6.4   Half-Space System for Coxeter Groups

Using the walls and roots introduced in Sect. 6.3 we are now ready to define a half-space system associated with a Coxeter group. The half-space system then gives rise to a CAT(0) cube complex on which the Coxeter group acts.

Definition 6.30 (Half-Spaces for Γ_W) Let (W, S) be a Coxeter system and let u, v be adjacent vertices in the Cayley graph Γ_W. That is, $v = us$ for some $s \in S$. We write

$$H(u, v) = \{w \in W \mid d_S(w, u) < d_S(w, v)\}$$

for the root determined by the ordered pair u, v. Define

$$\mathcal{H} := \{H(u, v) \mid u, v \in W, d_S(u, v) = 1\}$$

$$\star : \mathcal{H} \to \mathcal{H} : H(u, v) \mapsto H(v, u)$$

with $\mathcal{H}$ being ordered by inclusion, that is, we write

$$H(u, v) \leq H(w, x) \text{ if } H(u, v) \subset H(w, x).$$

The pair $(\mathcal{H}, \star)$ is called *half-space system of* (W, S). We sometimes refer to the roots $H(u, v)$ and $H(v, u)$ as half-spaces.

Every edge $\{u, v\}$ in the Cayley graph $\Gamma(W, S)$ of a Coxeter system defines two roots, sometimes also referred to as half-spaces, $H(u, v)$ and $H(v, u)$, and a wall M_r, where r is the reflection that satisfies $ru = v$. The wall M_r separates the two

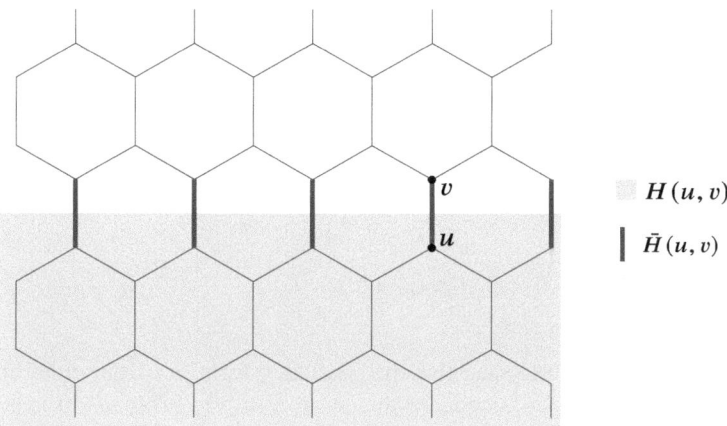

Fig. 6.6 The Cayley graph of the Coxeter group of type $\tilde{A}_2$ with one of its half-spaces $H(u, v)$ (shaded gray) and corresponding wall $\bar{H}(u, v)$ (bold red edges)

half-spaces in the sense that every edge-path connecting a vertex in one root with a vertex in the other root contains an edge in M_r.

We still have to prove that the half-space system of a Coxeter group indeed is a half-space system as defined earlier in this book. The roots play the role of the half-spaces. The wall M_r corresponds to the hyperplane in the half-space system defined by the pair of opposite roots (half-spaces).

Let us look at an example first.

Example 6.31 Let W be the Coxeter group of type $\tilde{A}_2$ given by the following presentation:

$$W = \langle s_0, s_1, s_2 \mid s_i^2, (s_i s_j)^3 \text{ for all } i, j \in \{0, 1, 2\} \rangle.$$

In Example 6.7 and Fig. 6.2 we have already encountered W and its Coxeter complex. The Cayley graph of W is the dual graph of the Coxeter complex. A vertex in the Cayley graph corresponds to a maximal cell in the Coxeter complex. Two vertices are connected if and only if the cells share a codimension-one face. Figure 6.6 shows the Cayley graph of this group.

A wall M_r of a reflection r is the set of all edges intersecting the reflection hyperplane. An example of such a wall is colored red in Fig. 6.6. The root (or half space) $H(u, v)$ determined by the edge $\{u, v\}$ contains all the vertices in the gray shaded region.

In the next proposition we show that the elements of $\mathcal{H}$ as in Definition 6.30 are rightfully called half-spaces. We essentially follow the exposition in [NR03]. However, the proof of Proposition 6.32 is new.

Proposition 6.32 (Half-Space System for Coxeter Groups) *The triple $(\mathcal{H}, \leq, \star)$ introduced in Definition 6.30 is a half-space system in the sense of Definition 5.10.*

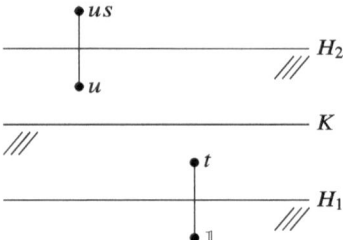

Fig. 6.7 Illustration of the proof of Proposition 6.32

Proof of Lemma 6.19 It is clear that the relation $\leq$ forms a partial order on the set of half-spaces. As a set W is equal to the union $H(u, v) \cup H(v, u)$. It is hence not hard to see that $\star$ is an order reversing involution of W. We are left with proving the following two statements:

- *Claim 1:* The nesting condition is satisfied, that is, at most one of the four conditions listed in Definition 5.10 holds.
- *Claim 2:* The finite interval condition is satisfied, that is, for all $h_1, h_2 \in \mathcal{H}$ there are only finitely many $k \in \mathcal{H}$ such that $h_1 \leq k \leq h_2$.

To prove Claim 1 check that the nesting condition is satisfied by any partial order defined by inclusion of a family of nonempty subsets of a set when the involution is given by the complement operator.

To prove that the finite interval condition holds we have to work a bit harder. Let $H_1 \subsetneq H_2$ be two half-spaces and suppose without loss of generality that

$$H_1 = H(\mathbb{1}, t) \text{ for some } t \in S \text{ and } H_2 = H(u, v) \text{ with } v = us, s \in S$$

as shown in Fig. 6.7. Otherwise move the H_i to such a position using the left-translation action of W on Γ_W. Lemma 6.29 implies that $t \notin H_1$, but $t \in H_2$ and $u \in H_2$, but $u \notin H_1$.

Now suppose that there exists a half-space K such that $H_1 \subsetneq K \subsetneq H_2$. Any such K satisfies $t \in K$ and $u \notin K$ by Lemma 6.29 and the fact that $K \neq H_i, i = 1, 2$. Suppose $\gamma = (\mathbb{1} = c_0, c_1, \ldots, c_l = u)$ is a minimal edge-path in the Cayley graph Γ_W connecting the identity $\mathbb{1}$ with the element u. Then γ crosses the wall of K exactly once (see Lemma 6.25), say at the index $1 \leq i \leq l - 1$. That is, $c_i \in K$ and $c_{i+1} \notin K$.

From Lemma 6.28 we may then deduce that $K = H(c_i, c_{i+1})$ and say that the wall of K is defined by γ. Moreover each such an edge-path $\gamma : \mathbb{1} \rightsquigarrow u$ defines at most $l - 1$ walls. Not all walls $H(c_i, c_{i+1})$ lie between H_1 and H_2, as illustrated in Fig. 6.8, but all that do are of this form.

Choose an element u' of minimal length that is contained in H_2 and for which $u's = v' \notin H_2$. Such a choice is possible since roots are convex subsets of the Cayley graph Γ_W with respect to $d = d_S$. Let v' in $-H_2$ be the vertex adjacent to u'

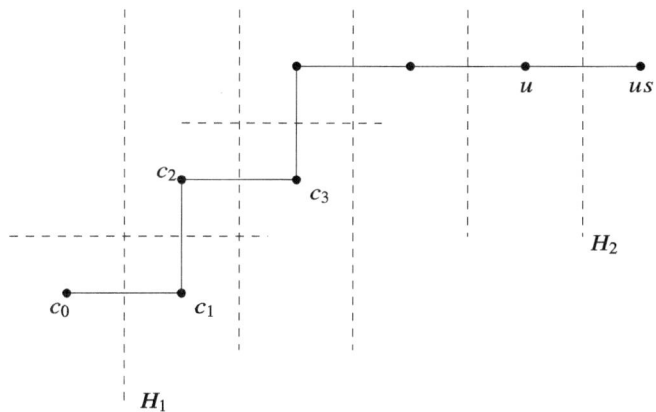

Fig. 6.8 Illustration of the proof of Proposition 6.32

with minimal distance to $\mathbb{1}$. (This vertex v' is the nearest point projection of $\mathbb{1}$ onto the convex subgraph of the Cayley graph spanned on the root $-H_2$.) But then, $\ell(u')$ is an upper bound on the number of possible walls between H_1 and H_2. In particular this number is finite.   □

Proposition 6.32 together with the cubulation construction explained in Chap. 5 imply that for any Coxeter group there exists a cube complex, obtained via cubulation of half space systems, on which the Coxeter group acts properly discontinuously.

Let G be a group acting on two different sets X and Y. Then a map $f : X \to Y$ is G-equivariant if $f(g.x) = g.f(x)$ for all $g \in G$, $x \in X$.

Proposition 6.33 (Cubulation of Coxeter Groups) *Suppose (W, S) is a Coxeter system, let Γ be its Cayley graph and let $X(\mathcal{H})$ be the cube complex defined in Definitions 5.16 and 5.20 associated with the half-space system introduced in Definition 6.30. Then*

1. *there exists a connected component Y of $X(\mathcal{H})$ admitting a G-equivariant embedding $\Phi : \Gamma \to Y$.*
2. *the action of W on Y is properly discontinuous.*

Proof of Lemma 6.19 Define a map $\Phi : \Gamma \to X(\mathcal{H})$ as follows: The set M_r of walls in Γ corresponds to pairs of half-spaces $H(u, v)$ and $H(v, u)$, where u, v are adjacent vertices in Γ. Write $\bar{H}(u, v)$ for such a pair of half-spaces and define for every $g \in W$ a map v_g by sending $\bar{H}(u, v)$ to the unique half-space in $\bar{H}(u, v)$ containing the element g as a vertex.

It is clear that for each $g \in \Gamma$ the map v_g defines a vertex in $X(\mathcal{H})$. The group W acts on $\mathcal{H}$ by $w.H(u, v) := H(w.u, w.v)$ for all $w \in W$. This induces a transitive action of W on the set $\{v_g \mid g \in W\}$ of vertex maps by putting $w.v_g := v_{w.g}$ for all w and all g.

It hence remains to show that adjacent vertices in Γ are mapped to adjacent vertices in $X(\mathcal{H})$. Observe that for any pair $g \in W$ and $s \in S$ the only wall $\bar{H}$ on which the maps v_g and v_{gs} differ is defined by the edge $\{g, gs\}$. Without loss of generality $g \in H$ and $gs \in H^\star$. Hence, Lemma 6.25 implies that $H = H(g, gs)$ and the claim follows.

We now need to show that vertex stabilizers $S_v := \mathrm{Stab}_W(v)$ are finite for arbitrary vertices $v \in Y$. By construction of the map Φ each vertex $v \in Y$ is at finite distance of $v_{\mathbb{1}} = \Phi(\mathbb{1})$, the image of the identity element. The orbit of $\mathbb{1}$ under S_v is the set of all vertices in Y having distance $d(v_i d, v)$ to v. In particular $S_v.v_i d$ is finite. The stabilizer S_v has a finite index subgroup, which is contained in $S_{v_i d}$, and hence S_v itself has to be finite as $S_{v_i d} = \{\mathbb{1}\}$.   □

By Proposition 6.33 every Coxeter group acts properly discontinuously on a cube complex. In order to apply the Švarc–Milnor theorem this action needs to be cocompact. However, this is not always the case as the following example shows.

Example 6.34 A Coxeter group W with a non-cocompact action on $X(\mathcal{H})$ is the group of type $\tilde{A}_2$, which we had already considered in Examples 6.31 and 6.7. Recall that this group is given by the presentation

$$ W = \langle r, s, t \mid s^2, t^2, r^2, (st)^3, (rs)^3, (rt)^3 \rangle. $$

Its Coxeter complex is isomorphic to the tiling of the Euclidean plane by equilateral triangles. The hyperplanes in the Cayley graph with respect to the standard generators correspond to the reflection hyperplanes inside the equilateral triangle tiling. The corresponding cube complex is the standard cubing of $\mathbb{R}^3$ and the quotient of the action is isomorphic to the real line. The three linearly independent directions in $\mathbb{R}^3$ correspond to the three parallel classes of hyperplanes in the triangle tiling of $\mathbb{R}^2$. Cubes correspond to triples of pairwise transversal hyperplanes. Compare Fig. 6.9 for a picture that illustrates this correspondence. The blue hexagons in this figure give rise to some of the 3-cubes in the cubulation. Exercise 5.47 walks you trough the steps to construct the cubulation.

The details of the cubulation of the type $\tilde{A}_2$ Coxeter group are worked out in [VSvRK21], where this cubulation is applied in a climate science context to compute connected components of cloud structures on the ICON grid.

Some Coxeter groups act cocompactly on their cubulation. An example is provided below.

Example 6.35 We apply the cubulation method to the group $G = PGL(2, \mathbb{Z})$ which is, by definition, the quotient of $PGL(2, \mathbb{Z})$ by the matrix $\left(\begin{smallmatrix} -1 & 0 \\ 0 & -1 \end{smallmatrix}\right)$ and is generated by the following three matrices:

$$ \begin{pmatrix} 0 & 1 \\ 1 & 0 \end{pmatrix}, \begin{pmatrix} -1 & 1 \\ 1 & 0 \end{pmatrix} \text{ and } \begin{pmatrix} 1 & 0 \\ 0 & -1 \end{pmatrix}. $$

Fig. 6.9 The hyperplanes in the equilateral triangulation of the plane are all parallel to one of the three colored, bold lines. The blue hexagons are part of the (dual) Cayley graph indicating some of the 3-cubes in the associated cubulation

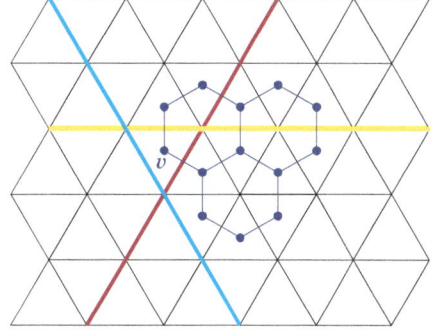

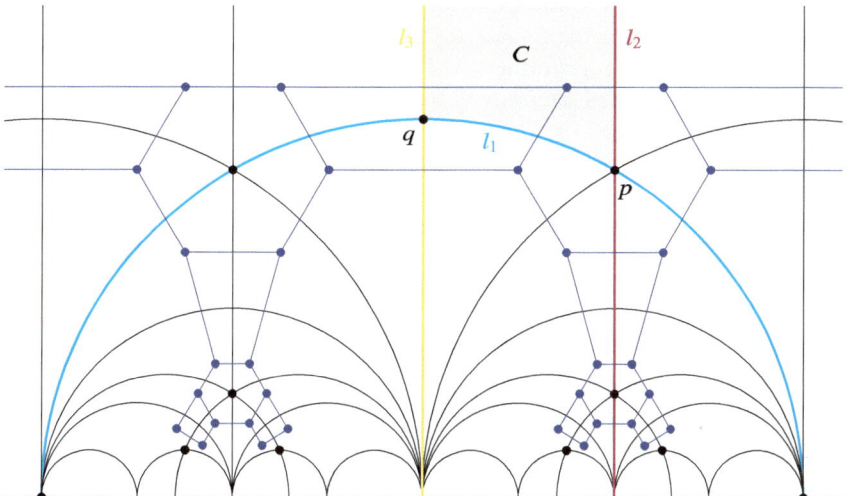

Fig. 6.10 Tiling of the hyperbolic plane induced by the action of $PGL(2, \mathbb{Z})$. The region labeled C is a fundamental domain for this action

As an abstract group G is isomorphic to the Coxeter group corresponding to the Coxeter diagram with three nodes where the first two are connected by an edge with no label and the second and third node are connected by an edge labeled ∞.

The group G acts by Möbius transformation on the upper half-plane model of $\mathbb{H}^2$. The action is defined as follows:

$$\begin{pmatrix} a & b \\ c & d \end{pmatrix} . z := \frac{az+b}{cz+d} \text{ for all } \begin{pmatrix} a & b \\ c & d \end{pmatrix} \in PGL(2, \mathbb{Z}).$$

A tiling of the hyperbolic plane preserved by this action is illustrated in Fig. 6.10. The three generators correspond to the reflections along the lines l_i, $i = 1, 2, 3$, shown in yellow, red and light blue. A fundamental domain of this action is labeled C and shaded gray.

The cubulation by half-space systems as introduced in Sect. 5.2 may be applied to the group $G = PGL(2, \mathbb{Z})$. One obtains a cube complex whose cubes correspond to families of pairwise intersecting hyperplanes in the hyperbolic plane when tiled as in Fig. 6.10. The group action induces a tiling of the hyperbolic plane. The dual graph of this tiling is shown in dark blue. This graph already resembles part of the one-skeleton of the cubulation. The hexagons correspond to 3-cubes and the 4-gons to 2-cubes in the cubulation.

A piece of the resulting cube complex is shown in Fig. 6.11 where it is also shown how the cubes fit together with the tiling of $\mathbb{H}^2$. The lines l_1, l_2 and l_3 in the tiling correspond to hyperplanes, also labeled l_i, in the cube complex. One can see in this figure that the action of G on the cube complex is in fact cocompact (look at the gray shaded region and its corresponding cubes in Fig. 6.11).

The hyperplanes of the half-space system are exactly the reflection hyperplanes of the action by Möbius transformation. There are two classes of such families. One is the class of hyperplanes intersecting in p or translates of p, which are shown in light blue and red in Fig. 6.11. The other is the class consisting of single edges, like the ones intersecting in q. The first class induces 3-cubes while the second induces 2-cubes in $X(\mathcal{H})$.

Cubes corresponding to adjacent vertices p, q in the hyperbolic plane share a face of dimension one in the cubulation. Compare the vertices labeled p, q in Fig. 6.10.

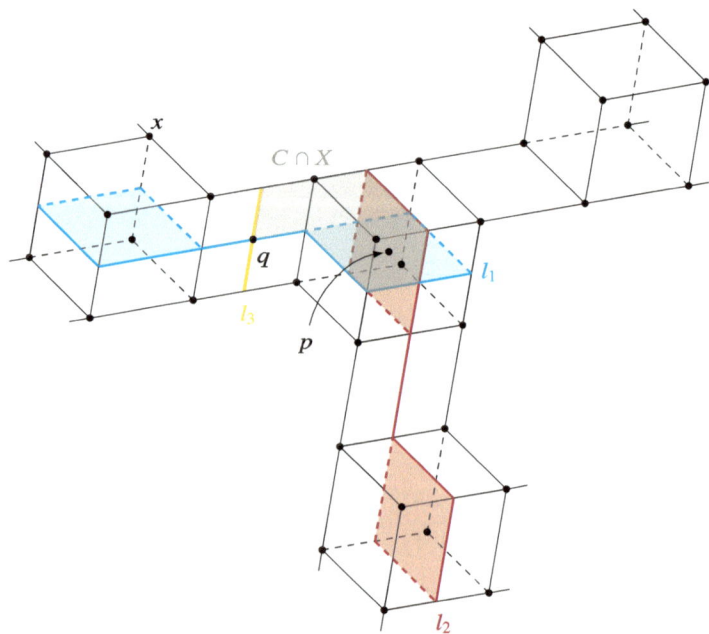

Fig. 6.11 There are two orbits of hyperplanes in the cubulation of $PGL(2, \mathbb{Z})$. This figure also shows the two classes of hyperplanes and (in gray) a fundamental domain for the action on the complex

These vertices correspond to centers of adjacent 2 and 3-dimensional cubes in Fig. 6.11.

For some classes of groups the action is in fact always cocompact.

Theorem 6.36 (Conditions for Cocompact Actions) *Let (W, S) be a Coxeter system. Then the action of W on its cube complex is cocompact if one of the following conditions is satisfied:*

1. *([CM05]) The Coxeter diagram of W does not contain a connected affine sub-diagram of rank ≥ 3.*
2. *([NR03]) W is right-angled.*
3. *([NR03]) W is word hyperbolic.*

We have seen that maximal cubes correspond to maximal families of pairwise transversal hyperplanes in the half-space system. In order to prove the first item one therefore has to show that there exists some $n \in \mathbb{N}$ such that every family of at least n hyperplanes contains at least one nested pair. See, for example, [NR03] for a proof.

6.5 Exercises

Exercise 6.37 Prove that the Coxeter matrix described in Example 6.5 indeed encodes the symmetric group $\mathrm{Sym}(m)$.

Exercise 6.38 Consider the following two Coxeter systems (W, S) and (W', S') with:

$$W = \langle s, t \mid s^2, t^2, (st)^6 \rangle$$

and

$$W' = \langle a, b, c \mid a^2, b^2, c^2, (ab)^3, (ac)^2, (bc)^2 \rangle.$$

These systems are also encoded by the following graphs:

- the graph of W has one edge with label 6. Vertices of this edge correspond to s and t.
- the graph of W' is not connected and consists of an edge (with label 3) with vertices a, b and one additional vertex c.

Work on the following:

1. Compute the set of involutions of W and W', i.e. the elements of order 2.
2. Compute for both W and W' the set of reflections, i.e. $R = \{wuw^{-1} \mid u \in S, w \in W\}$ and $R' = \{wuw^{-1} \mid u \in S', w \in W'\}$.

3. Prove that W and W' are isomorphic. Provide an explicit isomorphism $\phi : W \to W'$ (or in the other direction).
4. How do $\phi(R)$ and R' differ? What is the difference between R and the set of involutions?

Exercise 6.39 The Coxeter group of type A_n is defined to be the Coxeter group with generators $s_1, \ldots, s_n$ where s_j and s_k commute if $|j - k| \geq 2$ and $s_i s_{i+1}$ has order 3, for $1 \leq k \leq n - 1$.

1. Consider the symmetric group $Sym(n+1)$ of permutations of the set $\{1, \ldots, n+1\}$. If W is a Coxeter group of type A_n, show that there exists a surjective homomorphism $W \mapsto Sym(n+1)$ which maps s_k to the transposition $(k, k+1)$ for all $k \in \{1, 2, \ldots, n\}$.
2. Let $J = \{s_1, s_2, \ldots, s_{n-1}\}$. Show (for example by induction) that the cosets $W_J, s_n W_J, s_{n-1} s_n W_J, \ldots, s_1 \cdots s_n W_J$ exhaust all of W and hence W has cardinality at most $(n + 1)!$. Conclude that W is isomorphic to $Sym(n + 1)$.

Exercise 6.40 Let D_m be the dihedral group D_m. That is, D_m is generated by s and t such that their product has order m.

1. Prove that s and t are conjugate in D_m if and only if $m < \infty$ and m is odd.
2. Describe the center of the dihedral group D_m for all values of m including ∞.
3. Can you describe the conjugacy classes of all elements in D_m?

Exercise 6.41 Let $S = \{s_i, i = 1, \ldots, n\}$ be a set with n elements and let W be the Coxeter group generated by S with respect to some Coxeter matrix M. Consider the multiplicative group $Z_2 = \{-1, 1\}$ and let ϵ denote the homomorphism $\epsilon : W \to \{-1, 1\}$ defined by $\epsilon(s_i) = -1$ for all $s_i \in S$. Put $W^+ = ker(\epsilon)$ and prove the following: .

1. W^+ is generated by $\{s_i s_n | 1 \leq j \leq n - 1\}$.
2. If all entries $m_{i,j}$ of the Coxeter matrix M are even, then W has at least 2^n elements.

Exercise 6.42 Let W_1 and W_2 be two Coxeter groups given by the following presentations:

$$W_1 = \langle s, t \mid s^2, t^2, (st)^4 \rangle$$
$$W_2 = \langle r, s, t \mid r^2, t^2, s^2, (rs)^4, (st)^4, (rt)^2 \rangle$$

1. Construct the Coxeter complex $\Sigma_1 = \Sigma(W_1, \{s, t\})$.
2. Construct the Coxeter complex $\Sigma_2 = \Sigma(W_2, \{r, s, t\})$.
3. In how many ways can one embed Σ_1 into Σ_2?

Exercise 6.43 Let $W = \langle s, t \mid (st)^m, s^2, t^2 \rangle$ be a Coxeter group.

1. In case m is finite, determine the number of elements in W and prove that there is only one element with two reduced words representing it.

2. In case $m = \infty$ show that every element of W has a unique minimal presentation. How can one characterize these presentations?
3. Draw the Coxeter complex in case $m = 2, 3$ and ∞.

Exercise 6.44 (See Also Lemma 6.13) Let (W, S) be a Coxeter system and let $\ell \colon W \to \mathbb{N}_0$ be the word-length on W. Prove that either $l(ws) = l(w) + 1$ or $l(ws) = l(w) - 1$ holds for all $s \in S$ and $w \in W$.

Exercise 6.45 Prove that the conditions (D), (E) and (F) hold true for Coxeter groups and are equivalent.

Exercise 6.46 Let (W, S) be a Coxeter system. The wall H_r associated with a reflection r in W is defined to be the union of all simplices in $\Sigma = \Sigma(W, S)$ that are stable under the left-action of W on Σ.

1. Prove that walls are subcomplexes of the Coxeter complex of codimension one, which separate Σ in two disjoint subsets.
2. Describe the half-space system $\mathcal{H}'$ associated with walls as defined above.
3. Prove that $\mathcal{H}'$ is isomorphic to the half-space system $\mathcal{H}$ determined by the walls M_r as defined in Definition 6.21.

Chapter 7
A Panoramic Tour

This chapter highlights four different applications of and structural results about CAT(0) cube complexes. We discuss fixed point properties of actions on CAT(0) cube complexes in analogy to similar results about actions on trees in Sect. 7.1. To learn more about one of the many algebraic consequence for groups acting nicely on CAT(0) cube complexes, we prove in Sect. 7.2 that all such groups satisfy the Tits alternative.

Admitting an action on a special cube complex has even stronger algebraic consequences for a group. For example, all such groups embed into RAAGs. We introduce the class of special cube complexes and prove the group embedding property in Sect. 7.3. The material in this section should also be a good preparation to read about Agol's proof of the virtual Haken conjecture [Ago13]. Finally, in Sect. 7.4, we include an application outside pure mathematics and explain the construction of the space of phylogenetic trees. A proof that this space is a CAT(0) cube complex is provided.

7.1 Fixed Point Properties of Group Actions

Cube complexes are natural higher dimensional analogs of trees. Since groups acting on trees are known to have interesting fixed point properties it is only natural to ask whether similar properties hold true for cube complexes. In this section we consider isometric actions of finitely generated groups on cube complexes and study some of their fixed point properties. A crucial role is played by analogs of Helly's intersection theorem. We start with a recollection of the classical tree case.

7.1.1   Actions on Trees

Recall that a tree is a connected graph $\Gamma = (V, E)$ without circuits. By Exercise 3.43 their metric realizations are complete CAT(0) spaces. In this section we view trees as 1-dimensional cube complexes and study group actions on them.

First we require the group to respect the simplicial structure of the tree. More generally one can define:

Definition 7.1 (Polyhedral Actions) An isometric action $\phi : G \to \mathrm{Iso}(X)$ of a group G on a polyhedral (simplicial, cubical) complex X is *polyhedral (simplicial, cubical)* if it maps every cell (simplex, cube) of X to a cell (simplex, cube) of X.

In addition to study fixed points it is useful to avoid having edges setwise fixed without fixing their endpoints. This for example occurs when acting on a metric realization of a graph and reflecting an edge along its midpoint. We capture this behavior by the notion of inversions.

Definition 7.2 (Actions Without Inversions) An isometric action $\phi : G \to \mathrm{Iso}(\Gamma)$ of a group G on a graph Γ *has an inversion* if there exists some $g \in G$ and an edge $e = \{u, v\}$ in $E(\Gamma)$ such that $g.u = v$ and $g.v = u$. The element g is then called an *inversion*. We say that G acts *without inversion* on Γ if the action does not have an inversion.

Let X be the metric realization of a simplicial complex or cube complex. If an isometry $f : X \to X$ acts as an involution on an edge, then the end-vertices of the edge are swapped and the midpoint if fixed. In particular, inversions have fixed points. However, such a fixed point is not a cell of the underlying simplicial or cubical structure and is hence unwanted or artificial in some contexts. Therefore it is useful to assume that an action is simplicial and without inversions.

The guiding questions are the following.

Question 7.3 Suppose G is a finitely generated group. Does an isometric action of G on a tree T have a *global fixed point*? In other words, can we find a point x in T that is fixed by every element of the group G? What if we assume in addition that the action is without inversion?

It should be clear that the answer to this question depends on the group G. A first answer to Question 7.3 is provided by Helly's theorem for simplicial trees. Compare Lemma 10 in Serre's book [Ser03]. Before proving this result we first recall Helly's theorem in its classical form for subsets of $\mathbb{R}^n$.

Theorem 7.4 (Helly 1923) *Let S be a finite family $\{S_1, \ldots, S_l\}$ of subsets $S_i \subseteq \mathbb{R}^d$ where S_i is nonempty, closed and convex for $i = 1, \ldots, l$. If the intersection of any $(d + 1)$-tuple of elements in S intersects nontrivially, then $\cap S$ is also nonempty.*

Remark 7.5 The requirement that any $(d + 1)$ sets have nonempty intersection in Theorem 7.4 can not be weakened to d-tuples of sets. An example can easily be constructed in $\mathbb{R}^2$ where one can choose three pairwise intersecting lines that have an empty overall intersection.

A heuristic reason why Helly's theorem works is the specific behavior of convex sets in the flat plane. It is only natural to ask whether there is a generalization of this result in other situations or for other classes of non-positively curved spaces. A direct analog of Theorem 7.4 for trees is the following result:

Theorem 7.6 (Helly's Theorem for Trees) *Let T be a tree and $T_1, \ldots, T_k$ be connected subtrees of T. Then, the intersection of all subtrees is nonempty if the pairwise intersection of two subtrees is nonempty. I.e., if for all pairs of indices $1 \leq i, j \leq n$ one has $T_i \cap T_j \neq \emptyset$, then $\bigcap_{i=1}^{k} T_i \neq \emptyset$.*

The proof below is taken from [Led20, Lemma 3.20] expanding on [Ser03, Lemma 10].

Proof We prove the statement by induction on k. For $k \leq 2$ the assertion is clearly true. Let $k = 3$ and suppose for a contradiction that the intersection $T_1 \cap T_2 \cap T_3$ is empty. By assumption $T_i \cap T_j \neq \emptyset$. We can thus choose vertices $p_{i,j} \in T_i \cap T_j$ for all $i \neq j$ with $i, j \in \{1, 2, 3\}$ such that not all three are the same and such that the sum of their pairwise distances $d(p_{1,2}, p_{2,3}) + d(p_{1,3}, p_{2,3}) + d(p_{1,2}, p_{1,3})$ is minimal.

We now construct a cycle, i.e., a closed edge-path, in T. Let γ_i be the reduced path in T connecting $p_{i-1,i}$ with $p_{i,i+1}$ where we take indices mod 3 for $i = 1, 2, 3$. The concatenation c of the γ_i is by construction an edge-path from $p_{1,3}$ back to $p_{1,3}$ in T and hence a cycle. Since trees don't contain reduced cycles the edge-path c has to contain backtracking. Without loss of generality (up to switching indices) we may assume that there exists a vertex v such that $\gamma_1 = (p_{1,3}, \ldots, v, p_{1,2})$ and $\gamma_2 = (p_{1,2}, v, \ldots, p_{2,3})$.

Since sub-trees are connected and any two vertices in a tree are connected by a unique reduced path the edge-path γ_i is contained in T_i, for $i = 1, 2, 3$. But then, $v \in T_1 \cap T_2$ and

$$d(p_{1,3}, v) < d(p_{1,3}, p_{1,2}) \text{ and } d(p_{2,3}, v) < d(p_{2,3}, p_{1,2}),$$

which contradicts the minimality in the choice of the points $p_{i,j}$.

Thus, $T_1 \cap T_2 \cap T_3 \neq \emptyset$ and the assertion holds for $k = 3$.

Let $k \geq 3$ and suppose that the assertion is true for $n \leq k$ subtrees of T. Let $T_1, \ldots, T_{k+1}$ be sub-trees of T with pairwise non-empty intersections. Put $A_i = T_i$ for all $i = 1, \ldots, k - 1$ and put $A_k = T_k \cap T_{k+1}$. Then A_i with $i \leq k$ are trees. We claim that $A_i \cap A_j \neq \emptyset$. If $i, j < k$ then this is obviously true. So suppose that $i < k$ and $j = k$. Then $A_i \cap A_k = T_i \cap T_{k-1} \cap T_k$, which is non-empty, since the assertion holds for triples of sub-trees by the start of the induction. By the induction hypothesis we hence obtain $\bigcap_{i=1}^{k+1} T_i = \bigcap_{i=1}^{k} A_i \neq \emptyset$. □

We now look at group actions on trees.

Lemma 7.7 *Let G be a group acting without inversions on a tree T. Then for any* $g \in G$ *the fixed point set* $\text{Fix}_g(T)$ *is either empty or a connected subtree of T.*

Proof Suppose for a given $g \in G$ that the set of fixed points in nonempty, i.e.,

$$T^g := \text{Fix}_g(T) \neq \emptyset.$$

Since G acts without inversions, the set T^g is a subgraph of T and hence either a tree or a union of trees. It remains to prove that T^g is connected. Let v, v' be two distinct vertices in T^g and let $c = (v = v_0, v_1, \ldots, v_{n-1}, v_n = v')$ be a reduced path from v to v'. Its image $g(c) = (v = g(v_0), g(v_1), \ldots, g(v_{n-1}), g(v_n) = v')$ is also reduced. Since reduced paths in trees are unique $c = g(c)$ and T^g is connected.   □

A consequence of Theorem 7.6 is the next corollary.

Corollary 7.8 *Let G be a finitely generated group with generating set* $\{g_1, \ldots, g_k\}$ *and suppose there exists a simplicial action* Φ *of G on a tree T, which is without inversions. Suppose that* $\text{Fix}_\Phi(\langle g_i \rangle) \cap \text{Fix}_\Phi(\langle g_j \rangle) \neq \emptyset$ *for all* $i, j \in \{1, \ldots, k\}$. *Then G has a global fixed point, i.e.,* $\text{Fix}_\Phi(G) \neq \emptyset$.

Proof The fixed point sets $\text{Fix}_\Phi(\langle g_i \rangle) \subseteq X$ are non-empty subtrees of T by Lemma 7.7. We may thus apply Helly's theorem for trees, Theorem 7.6, and obtain that

$$\emptyset \neq \bigcap_{i=1}^{k} \text{Fix}_\Phi(\langle g_i \rangle) = \text{Fix}_\Phi(\langle g_1, \ldots, g_k \rangle) = \text{Fix}_\Phi(G)$$

hence there exists a global fixed point.   □

A slight generalization is possible, namely the statement also holds without the assumption that the action is without inversions. See Exercise 7.95.

We illustrate this result with an example, which is taken from [Bri08]. In this paper Bridson shows that every action of $\text{Aut}(F_n)$ on an $\mathbb{R}$-tree has a fixed point. Fixed point properties for actions on higher dimensional CAT(0) spaces are studied in [Bri07]. Varghese generalized this result in [Var19] to strongly simplicial actions on CAT(0) cube complexes. She shows that such an action has a global fixed point. In particular, any such action of $\text{Aut}(F_n)$ has a fixed point. See also [Var14].

Example 7.9 ([Bri08]) Let $F_n = \langle x_1, \ldots, x_n \rangle$ be the free group of rank n for some $n \geq 3$ and write $\text{Aut}(F_n)$ for the automorphism group of F_n. The composition of two automorphisms $\alpha, \beta \in \text{Aut}(F_n)$ is given by $\alpha\beta := \beta \circ \alpha$.

A *right-Nielsen automorphism* is defined as follows: for $i, j \in \{1, \ldots, n\}, i \neq j$ put

$$\rho_{i,j} : F_n \longrightarrow F_n$$

$$x_k \longmapsto \begin{cases} x_i x_j, & \text{if } k = i \\ x_k, & \text{if } k \neq i. \end{cases}$$

Involutions $(x_i, x_j) : F_n \longrightarrow F_n$ are defined by

$$(x_i, x_j)(x_k) := \begin{cases} x_j, & \text{if } k = i \\ x_i, & \text{if } k = j \\ x_k, & \text{else} \end{cases}$$

and automorphisms $e_i : F_n \longrightarrow F_n$ by putting

$$e_i(x_k) := \begin{cases} x_i^{-1}, & \text{if } k = i \\ x_k, & \text{if } k \neq i. \end{cases}$$

By construction $\rho_{i,j}$, (x_i, x_j) and e_i are elements of $\mathrm{Aut}(F_n)$ and one can show that in fact $\mathrm{Aut}(F_n)$ is generated by the set

$$Y := \{(x_1, x_2)e_1 e_2, (x_2, x_3)e_1, (x_i, x_{i+1}), e_2 \rho_{1,2}, e_n \mid i = 3, \ldots, n - 1\}.$$

Observe that for all $\alpha \in Y$ the order $\mathrm{ord}(\alpha)$ of α is two and that $|\langle \alpha, \beta \rangle| < \infty$ for any pair $\alpha, \beta \in Y$.

Let now X be a tree and $\Phi : \mathrm{Aut}(F_n) \to \mathrm{Iso}(X)$ an isometric action of $\mathrm{Aut}(F_n)$ on this tree. Then this action has a fixed point. To see this argue as follows.

The Bruhat-Tits fixed point Theorem 3.15 implies that $\mathrm{Fix}_\Phi(\langle \alpha \rangle) \neq \emptyset$ as trees are CAT(0) and complete. Moreover $\mathrm{Fix}_\Phi(\langle \alpha \rangle) \cap \mathrm{Fix}_\Phi(\langle \beta \rangle) \neq \emptyset$ for arbitrary $\alpha, \beta \in Y$. Now using Exercise 7.95 we may conclude that $\mathrm{Fix}_\Phi(\mathrm{Aut}(F_n)) \neq \emptyset$.

The property that a certain group always has to have a fixed point when acting on a space of a certain class is widely studied. For trees this notion was introduced by Serre and reads as follows.

Definition 7.10 (Property FA**)** A group G has *property* FA if every isometric action without inversion of G on a tree T has a global fixed point, that is, there exists a vertex x such that $g(x) = x$ for all $g \in G$.

The name FA comes from the French expression *fixe arbre*, which translates as fixing a tree.

Example 7.11

1. One can show that finite groups and also more generally finitely generated torsion groups have property FA. (See [Ser03] Exercise 6.3.1 and Theorem 15).
2. We have seen in Example 7.9 that the group $\mathrm{Aut}(F_n)$ has property FA for all $n \geq 3$.
3. Also $GL_m(\mathbb{Z})$ has FA for $m \geq 3$, as shown in [Vog02, p.23].
4. It is known that finite index subgroups of $\mathrm{Aut}(F_3)$ do not have property FA. On the contrary $\mathrm{Aut}(F_n)$ for $n \geq 5$ has property (T) and hence itself and all its finite index subgroups have property FA.
5. The action of $\mathbb{Z}$ on the real line is fixed point free and without inversions. Hence $\mathbb{Z}$ does not have property FA.

Apart from being isomorphic to $\mathbb{Z}$ there is another obstruction to property FA. Amalgamated products and HNN extensions also do not have FA. This is shown in [Ser03] Theorem 15. The proof uses Bass-Serre-trees for the group. See Sect. 7.2.

Theorem 7.12 (Serre's Theorem) *If a finitely generated group G has property* FA *then G can not be written as nontrivial amalgamated product.*

We aim to generalize property FA to cube complexes and hence introduce the following definition.

Definition 7.13 (Property F$\mathcal{X}$) Let $\mathcal{X}$ be a class of metric spaces. We say that a group G has property F$\mathcal{X}$ if every isometric action of G on a space $X \in \mathcal{X}$ has a global fixed point.

7.1.2   Helly's Theorem for CAT(0) Cube Complexes

The goal for this subsection is to prove Theorem 7.18, an analog of Helly's theorem for CAT(0) cube complexes. Recall that a cube subcomplex Y of a cube complex X is a subset consisting of cubes of X with cubical structure (and gluing) inherited from X. See Definition 4.9 for a formal definition. We do not assume that subcomplexes are connected. Compare Fig. 7.1.

We collect some tools needed for the proof of Theorem 7.18.

Fig. 7.1 Y embeds as a subcomplex into X in many ways

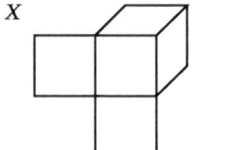 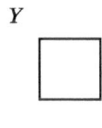

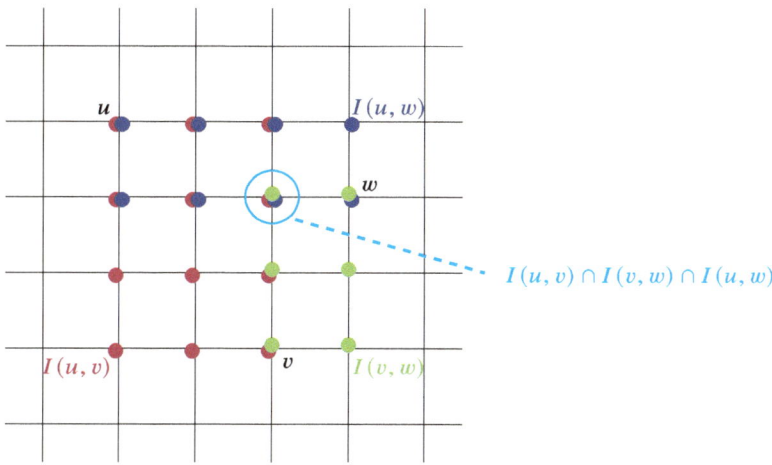

Fig. 7.2 A median graph with three intervals intersecting in a unique (circled) vertex

Definition 7.14 (Median Graph) Let $\Gamma = (V, E)$ be a graph with distance function d (as defined in Definition 4.26) on its vertices. The *interval* $I(u, v)$ between vertices $u, v \in V$ in Γ is the set

$$I(u, v) := \{x \in V : d(u, v) = d(u, x) + d(x, v)\}.$$

We call Γ *median*, if for pairwise different $u, v, w \in V$ there exists a unique vertex x in the intersection $I(u, v) \cap I(u, w) \cap I(v, w)$.

Median graphs are in close connection with CAT(0) cube complexes as shown in Theorem 7.17.

Example 7.15 The one-skeleton of the standard cubing of the Euclidean plane is median. See Fig. 7.2, where a piece of the graph and three intervals are shown (in blue, orange and green). The unique vertex in their intersection is circled.

In CAT(0) cube complexes one has a nice characterization of points in intervals.

Lemma 7.16 *Let X be a CAT(0) cube complex and let $x, y \in X^{(0)}$ be two vertices in X. Then $z \in I(x, y)$ if and only if no hyperplane separates z from the two vertices x and y.*

Proof Let $W(x, y)$ be the set of hyperplanes in X such that H separates x and y. Exercise 7.96 implies that the number of elements in $W(x, y)$ equals $d(x, y)$. A vertex z is in the interval of x and y if, by definition, $d(x, y) = d(x, z) + d(z, y)$. This is in turn equivalent to the fact that the number of elements in the set $W(x, y)$ is the same as the number of elements in $W(x, z)$ plus the number of elements in $W(z, y)$. But, the sizes of the sets $W(x, z)$ and $W(z, y)$ add up to the size of $W(x, y)$ if and only if $W(x, y) = W(x, z) \sqcup W(z, y)$. There is thus no hyperplane

Fig. 7.3 Illustration of a step
in the proof of Theorem 7.17

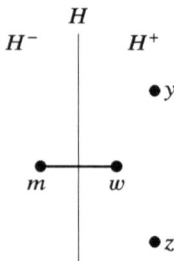

simultaneously contained in $W(x, z)$ and $W(y, z)$. Hence, no hyperplane separating z from x and y. If conversely there existed such a hyperplane, the set would not be a disjoint union.   □

In Example 7.15 we have seen a CAT(0) cube complex whose one-skeleton is median. The next theorem shows that this is no coincidence.

Theorem 7.17 (CAT(0) **Implies Median**) *The one-skeleton of every* CAT(0) *cube complex is a median graph.*

Proof Suppose X is a CAT(0) cube complex and fix pairwise distinct $x, y, z \in X^{(0)}$.

We first show that $I(x, y) \cap I(x, z) \cap I(y, z) \neq \emptyset$. Choose a vertex $m \in I(x, y) \cap I(x, z)$ such that $d(x, m)$ is maximal. We need to prove that $m \in I(y, z)$.

Suppose for a contradiction that $m \notin I(y, z)$. Lemma 7.16 implies that there exists a hyperplane H separating m from y and z. Suppose without loss of generality that H has minimal distance to m with this property. Call H^- the half-space determined by H that contains m and refer to the other half-space by H^+. By construction H^+ contains y and z.

Let $w \in H^+$ such that $d(m, w) = 1$. Compare Fig. 7.3 for an illustration of the current situation.

We claim that $w \in I(x, y)$. Suppose otherwise. Then there exists, by Lemma 7.16, a hyperplane J separating w from x and y. Let J^+ be the half-space determined by J that contains w and refer to the other half-space by J^-. By construction J^- contains x and y. The point m is not in J^+ since $m \in I(x, y)$ and hence in J^-. This implies that J separates w and m. But, the distance $d(m, w) = 1$ and w and m are connected by an edge which spans a unique hyperplane. This implies that $J = H$. But, H separates m and y. This leads to a contradiction.

Swapping y for z we obtain that $w \in I(x, z)$. Further

$$d(x, w) = d(x, m) + d(m, w) \geq d(x, m).$$

But, this contradicts the choice of m, which was a vertex in the intersection $I(x, y) \cap I(x, z)$ with maximal distance to x. Hence

$$I(x, y) \cap I(x, z) \cap I(y, z) \neq \emptyset.$$

To prove uniqueness we proceed as follows. Let $\mathcal{H}$ be the collection of half-spaces containing at least two of the three vertices x, y, z. Lemma 7.16 implies that the interval $I(x, y)$ is contained in the intersection of all half-spaces containing x and y. Therefore

$$I(x, y) \cap I(x, z) \cap I(y, z) \subseteq \bigcap \mathcal{H}.$$

Suppose there are two distinct vertices $m_1, m_2 \in I(x, y) \cap I(x, z) \cap I(y, z)$. Then there exists at least one hyperplane separating m_1 and m_2 and $\#W(m_1, m_2) = d(m_1, m_2) > 0$. Let H be such a hyperplane separating m_1 and m_2 and assume without loss of generality that x, y are contained in a common half-space determined by H. Call this half-space H^- and observe that $H^- \in \mathcal{H}$. Its complement H^+ contains one of the two vertices m_i, say m_2, and is not contained in $\mathcal{H}$.

Since the set $I(x, y) \cap I(x, z) \cap I(y, z)$ is contained in the intersection of all hyperplanes in $\mathcal{H}$ we obtain that $m_2 \notin \bigcap \mathcal{H}$ and hence also $m_2 \notin I(x, y) \cap I(x, z) \cap I(y, z)$. This contradicts the choice of m_2 and uniqueness follows.   □

In fact, the converse of Theorem 7.17 is also true and every median graph is the one-skeleton of a CAT(0) cube complex. The equivalence of these notions is shown by Chepoi, see Theorem 6.1 in [Che00].

We are now ready to state and prove the main result of this subsection.

Theorem 7.18 (Helly's Theorem for CAT(0) **Cube Complexes)** *Let X be a CAT(0) cube complex and $\mathcal{S}$ a finite family of nonempty, convex subcomplexes. If the pairwise intersection of elements of $\mathcal{S}$ is nonempty, then $\bigcap_{S \in \mathcal{S}} S \neq \emptyset$.*

Proof Let $\mathcal{S} = \{X_1, \ldots, X_n\}$ for some $n \in \mathbb{N}$. We prove by induction that k-element subsets of $\mathcal{S}$ have nontrivial intersections. For $k = 2$ this claim follows directly from the assumptions. Put $Y := X_3 \cap X_4 \cap \cdots \cap X_k$. We know from the induction hypothesis that $X_1 \cap X_2 \neq \emptyset$, $X_1 \cap Y \neq \emptyset$ and $X_2 \cap Y \neq \emptyset$.

We also know that the sets Y, $X_1 \cap X_2$, $X_1 \cap Y$ and $X_2 \cap Y$ are convex and thus CAT(0) subcomplexes of X. Fix vertices $p \in X_1 \cap Y$, $q \in X_2 \cap Y$ and $r \in X_1 \cap X_2$. As Y is a CAT(0) subcomplex of X and $p, q \in Y$ the interval $I(p, q)$ is also contained in Y. Similarly, as p and r are contained in X_1 we obtain that also $I(p, r) \subseteq X_1$ and since q and r are in X_2 we also conclude that $I(q, r) \subseteq X_2$. Moreover we obtain

$$I(p, q) \cap I(p, r) \cap I(q, r) = \{m(p, q, r)\}$$

and $m(p, q, r) \in Y$. Hence, $m(p, q, r) \in X_1 \cap X_2 \cap Y$. And the Lemma follows.   □

Note that Helly's theorem holds true for general CAT(0) metric spaces. Several slightly different versions are known one of which is the following. Compare, for example, [Kle99] Theorem 5.3, [Far09] Theorem 3.2, [Bri12] Theorem 3.2 or the Theorem on p. 240 in [Var14].

Theorem 7.19 (Helly's Theorem for CAT(0) **Metric Spaces)** *Let X be a complete* CAT(0) *metric space of covering dimension d. Suppose $\mathcal{S}$ is a finite family of nonempty, closed and convex subsets. If every $(d+1)$-tuple of elements in $\mathcal{S}$ has a nonempty intersection, then $\bigcap_{S \in \mathcal{S}} S \neq \emptyset$.*

Apply Theorem 7.18 to the collection $\mathcal{S} := \{\text{Fix}(\langle g_1 \rangle), \ldots, \text{Fix}(\langle g_k \rangle)\}$ of fixed point sets of an isometric action and obtain the following corollary as a direct consequence.

Corollary 7.20 *Suppose G is a finitely generated group and let $\{g_1, g_2, \ldots, g_k\}$ be a set of generators. Let X be a* CAT(0) *cube complex on which G acts isometrically. If the fixed point sets* $\text{Fix}(\langle g_i \rangle)$ *are cube subcomplexes for all i and if*

$$\text{Fix}(\langle g_i \rangle) \cap \text{Fix}(\langle g_j \rangle) \neq \emptyset \text{ for all } i, j \in \{1, \ldots, k\},$$

then $\text{Fix}(G) \neq \emptyset$ *and the action has a global fixed point.*

Remark 7.21 The extra assumption that the fixed point-sets of generators of G are cube subcomplexes is needed. In general, such fixed point sets need not be subcomplexes. To see an example put $G = \mathbb{Z}/2\mathbb{Z}$ and let $X = \mathbb{R}^2$. Then define an action of G on X by letting $\bar{0} \in G$ act as the identity on $\mathbb{R}^2$ and by letting $\bar{1} \in G$ act as the reflection represented by the matrix $\begin{pmatrix} 0 & 1 \\ 1 & 0 \end{pmatrix}$. The fixed point set of the action of G is that of $\bar{1}$ as illustrated in Fig. 7.4. This diagonal (shown in blue) is clearly not a cube sub-complex.

We conclude this section with a fixed point criterion due to Varghese [Var19], which is closely related to Corollary 7.20.

A direct consequence of the Bruhat-Tits fixed point Theorem 3.15 is that a finite group can not act without fixed point on a CAT(0) cube complex.

Lemma 7.22 *Every isometric action of a finite group on a complete* CAT(0) *cube complex has a fixed point.*

Fig. 7.4 Fixed point set of a $\mathbb{Z}/2\mathbb{Z}$ action on R^2

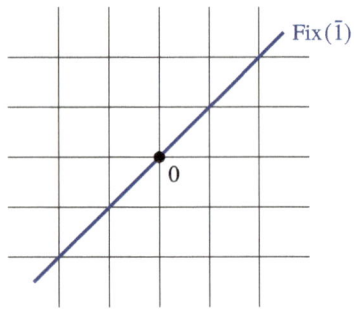

Using this lemma one can prove a fixed point criterion for actions satisfying an extra assumption. The notion of a strongly simplicial action can be seen as a higher dimensional analog of the notion of an action without inversions on a graph.

Definition 7.23 We say that an action of a group G on a cube complex is *strongly simplicial* if the stabilizer of any cube fixes that cube pointwise.

Theorem 7.24 (Fixed Point Criterion) *Let G be a group finitely generated by a set $S \subset G$. Suppose that each pair of elements in S generates a finite subgroup of G. Then every strongly simplicial action of G on a finite dimensional* CAT(0) *cube complex has a fixed point.*

Proof Lemma 7.22 implies that $\mathrm{Fix}(\langle s_1, s_2 \rangle)$ is nonempty for every pair of distinct generators $s_1, s_2 \in S$. As $\mathrm{Fix}(\langle s_1, s_2 \rangle) = \mathrm{Fix}(s_1) \cap \mathrm{Fix}(s_2)$ we have that every generator has a nonempty fixed point set. The assertion follows from Theorem 7.18 applied to the collection $\mathcal{S} := \{\mathrm{Fix}(\langle s_1 \rangle), \ldots, \mathrm{Fix}(\langle s_k \rangle)\}$ of fixed point sets.   □

As a consequence of this fixed point criterion Varghese obtains that for all $n \geq 3$ the group $\mathrm{Aut}(F_n)$ can not act strongly simplicial and without a global fixedpoint on a finite dimensional CAT(0) cube complex. See [Var19] Theorem A.

7.2   The Tits Alternative

Groups acting cocompactly on a CAT(0) cube complex have many interesting properties. They are, for example, bi-automatic [Wis04]. Another consequence is that such a group satisfies the Tits alternative. That is, every subgroup is either virtually solvable or contains a non-abelian free subgroup.

In this chapter we prove a version of the Tits alternative for groups G acting geometrically on a CAT(0) cube complex. This result, as stated in Theorem 7.55, is due to [SW05].

Recall that a group H is *virtually* P for some property P if H contains a finite index subgroup that is P.

Definition 7.25 (The Tits Alternative) A group G satisfies the *Tits alternative* if every subgroup H is either virtually solvable or contains a non-abelian free subgroup.

Remark 7.26 The first to prove such a property was Tits who showed in [Tit72] that finitely generated linear groups satisfy this dichotomy. Hence the name. Other prominent examples include hyperbolic groups [Gro87, GdlH90], mapping class groups [McC85, Iva84], the group $\mathrm{Out}(F_n)$ [BFH00] and certain classes of groups acting on nice complexes [OPM21].

Closely related to the results in this section is the work of Ballmann and Świątkowski [BŚ99] and Xie [Xie04]. In case of Artin groups of FC type the Tits alternative was shown by Martin and Przytycki [MP20] using CAT(0) cube complexes.

Versions of the Tits alternative similar to the one considered in this section have been shown by Caprace and Sageev in [CS11]. Their proofs rely on the dynamics of the group action on the boundary and on their rank rigidity result for CAT(0) cube complexes. Fioravanti [Fio18] extended their methods to median spaces by using the action on the Roller boundary of a median space.

In full generality, i.e. for general CAT(0) spaces, both rank rigidity and the Tits alternative are open problems to date.

Examples of groups not satisfying the Tits alternative include many big mapping class groups [All21], some Burnside groups [dlH00] and Thompson's group F [BS85]. We elaborate a bit on the latter in the next example.

Example 7.27 Thompson's group F is the group given by the following presentation:

$$F = \langle A, B \mid [AB^{-1}, A^{-1}BA] = [AB^{-1}, A^{-2}BA^2] = 1 \rangle.$$

Despite the innocent appearance of F this group is an important (counter-)example in geometric group theory, which has many interesting properties. See, for instance, the introduction by Cannon et al. [CFP96] for an overview.

In preparation for the proof of Theorem 7.55 we introduce three tools and concepts: HNN-extensions and free amalgamations, (relative) ends of groups and Fuchsian groups.

7.2.1   HNN-Extensions and Free Amalgamations

We introduce constructions to produce new groups out of old ones. First we define free amalgamated products of pairs of groups G_1, G_2, which share a subgroup. This constructions gives us a group G with the property that both G_i naturally embed in G and the isomorphic subgroups are conjugate in G.

Definition 7.28 (Amalgamated Free Product) Let $G_i = \langle S_i \mid R_i \rangle$, $i = 1, 2$ be two groups and suppose there exist isomorphic subgroups $H_i \leq G_i, i = 1, 2$. Fix an isomorphism $\phi : H_1 \to H_2$. The *amalgamated free product* of G_1 and G_2 over $H \cong H_1 \cong H_2$ (with respect to ϕ) is then defined by

$$G_1 \star_H G_2 := \langle S_1 \sqcup S_2 \mid R_1 \sqcup R_2 \sqcup \{\phi(h)h^{-1} \mid h \in H_1\} \rangle.$$

HNN-extensions mimic amalgamated products, but with two isomorphic subgroups within a single group. The main feature of these constructions is that a given group G is embedded in a larger group G' such that its isomorphic subgroups are conjugate in G' through a fixed isomorphism. The name HNN-extension refers to Graham Higman and Bernhard and Hanna Neumann who first introduced this concept [HNN50].

Definition 7.29 (HNN-Extension) Let $G = \langle S \mid R \rangle$ be a finitely presented group. Suppose H_1 and H_2 are two isomorphic subgroups of G. Fix an isomorphism $\phi : H_1 \to H_2$. The *HNN-extension* (with respect to ϕ) of G by a stable generator $t \notin G$ conjugating H_1 with H_2 is defined by

$$G \star_H := \langle S \sqcup \{t\} \mid R \sqcup \{tht^{-1} = \phi(h) \mid h \in H_1\}\rangle.$$

Remark 7.30 Despite the fact that both HNN-extensions and amalgamated free products depend on the chosen isomorphism $\phi : H_1 \to H_2$ this is not reflected in the notation. One typically writes $G \star_H$, respectively $G_1 \star_H G_2$, and suppresses the choice of the isomorphism (and the subgroups in the second case) in the notation.

A common generalization of free amalgamations and HNN-extensions is the concept of a graph of groups. See [BH99] for more on this topic. We now dip our toes into Bass-Serre theory and define trees associated with the two constructions.

Definition 7.31 (Bass–Serre Tree) The *Bass–Serre tree* associated with a free amalgamated product $G = A \star_C B$ is a quotient of the disjoint union of copies of $[0, 1]$ indexed by elements of G, which is defined as follows:

$$T := (G \times [0, 1]) / \sim$$

where the equivalence relation $\sim$ is induced by the three relations

$$(ga, 0) \sim (g, 0), (gb, 1) \sim (g, 1) \text{ and } (gh, t) \sim (g, t)$$

for all group elements $g \in G$, $a \in A$, $b \in B$ and $h \in C$ and parameters $t \in [0, 1]$. The images of the endpoints $0, 1$ of these intervals in the quotient T are the *vertices* of T and the images of the intervals themselves are the *edges* of T. The left-translation of G on $G \times [0, 1]$ is compatible with these relations, which in particular means that G acts on T by isometries.

By construction the Bass–Serre tree defined above (and also the one below) is a metric space which is the geometric realization of a simplicial tree. Compare also [BH99, p.355 and Thm. II.11.18].

Definition 7.32 (Bass–Serre Tree) The *Bass-Serre tree* associated with an HNN-extension $G = A \star_C$ is defined as follows:

$$T := (G \times [0, 1]) / \sim$$

where the equivalence relation $\sim$ is induced by

$$(g, s) \sim (g\phi(h), s), (g, 0) \sim (ga, 0) \sim (gt, 1)$$

for all $g \in G$, $a \in A$, $h \in C$ and $s \in [0, 1]$ and with t being the conjugating parameter from Definition 7.29. The images of the endpoints $0, 1$ of these intervals in the quotient T are the *vertices* of T and the images of the intervals themselves are the *edges* of T.

Again G acts on T by isometries and the quotient of this action is a complex consisting of a single vertex and an edge with both ends glued to that vertex. The stabilizer of every edge in T is isomorphic to C, the stabilizer of every vertex is isomorphic to A. The number of edges connected to a same vertex in T equals the index $[A : C]$.

We use normal forms to represent elements in HNN-extensions and free amalgamated products.

Definition 7.33 (Reduced Expression (HNN-Extension)) Let $A = \langle S | R \rangle$ be a finitely presented group, let C_i, $i = 1, 2$, be isomorphic subgroups. Suppose C is such that $C \cong C_1 \cong C_2$, and let $\phi : C_1 \to C_2$ be an isomorphism. Write G for the HNN-extension $G = A \star_C$. An expression $a_0 t^{m_1} a_1 t^{m_2} a_2 \cdots t^{m_n} a_n$ is *reduced* if it does not contain a consecutive sub-expression of the following form:

1. $t^{m_i} a_i t^{m_{i+1}}$, with $a_i \in C_1$ and exponents $m_i < 0$ and $m_{i+1} > 0$
2. $t^{m_j} a_j t^{m_{j+1}}$, with $a_j \in C_2$ and exponents $m_j > 0$ and $m_{j+1} < 0$.

Two distinct reduced expressions may be equal as elements in G. To obtain a normal form we have to refine the definition.

Definition 7.34 (Normal Form (HNN-Extension)) A *normal form* of an element g in an HNN-extension $G = A \star_C$ is a word $a_0 t^{m_1} a_1 t^{m_2} a_2 \cdots t^{m_n} a_n = g$ where a_0 is an arbitrary element in A and

1. if $m_i = -1$, then a_i is a coset representative of C_1 in G,
2. if $m_i = 1$, then a_i is a coset representative of C_2 in G, and
3. there is no sub-expression $t^m \mathbb{1} t^{-m}$.

Higmann, Neumann and Neumann proved that A is embedded in $G = A \star_C$. Combining this with a result of Britton one obtains the first statement in the normal form theorem.

Theorem 7.35 (Normal Form Theorem for HNN-Extensions) *Suppose* $G = A \star_C$ *as in Definition 7.33. Then the following two equivalent statements hold:*

1. *The group A is embedded in G via the identity map. The expression consisting of only g is a normal form. If $a_0 t^{m_1} a_1 t^{m_2} a_2 \cdots t^{m_n} a_n = \mathbb{1}$ in G, then it is not reduced.*
2. *Every element g of G has a unique normal form*

$$ g = a_0 t^{m_1} a_1 t^{m_2} a_2 \cdots t^{m_n} a_n. $$

A proof of this theorem can be found on page 182 in [LS01]. We follow their exposition and use this theorem to prove a normal form theorem for free products with amalgamation. See Theorem 2.6 on page 187 in [LS01].

Definition 7.36 (Reduced Expression (Amalgamated Free Products)) Suppose
$G = A \star_C B$. A sequence of elements $g_0, g_1, \ldots, g_n$ of G with $n \geq 0$ is *reduced*
if

1. $g_i \notin C$ for all $i > 1$.
2. g_i is in A or B for all i and successive g_i, g_{i+1} for $i > 0$ are never contained in
   the same factor. I.e. the g_i, for $i > 0$, alternate between $A \setminus C$ and $B \setminus C$.
3. If $n = 0$ the element $g_0 \neq \mathbb{1}$.

Clearly every nontrivial element in g is the product of the elements in a reduced
sequence. Conversely one can show

Theorem 7.37 (Normal Form Theorem for Amalgamated Free Products) *Suppose $G = A \star_C B$. If $g_0, g_1, \ldots, g_n$ is a reduced sequence in G, then*

$$g_0 g_1 \cdots g_n \neq \mathbb{1} \text{ in } G.$$

The natural homomorphisms $A \to A\star_C$ and $A, B \to A \star_C B$ are injective.

Proof The group G is given by the presentation

$$\langle A, B \mid c = \phi(c) \text{ for all } c \in C \rangle.$$

Define a new group F by putting

$$F = \langle A, B, h \mid t^{-1} ct = \phi(c) \text{ for all } c \in C \rangle.$$

There is a homomorphism $\Psi : G \to F$ defined by $\Psi(a) = t^{-1} at$ for all $g \in A$
and $\Psi(b) = b$ for all $b \in B$. The defining relations of G are mapped onto the
identity. Hence, this map is a homomorphism. If $n = 0$ and $\mathbb{1} \neq g_0 = a \in A$ then
$\Psi(a) = t^{-1} at = \phi(a) \neq \mathbb{1}$. In all other cases every reduced sequence is mapped to
a reduced sequence of F. Hence the result follows from the Normal Form Theorem
for HNN-extensions.   □

Remark 7.38 From Theorems 7.37 and 7.35 one can deduce:

1. $A\star_C$ contains a subgroup isomorphic to F_2 if $C_1 \neq A$ and $C_2 \neq A$.
2. $A \star_C B$ contains a subgroup isomorphic F_2 if $[A : C] > 2$ and $[B : C] > 2$.

7.2.2 Ends of Groups

The number of ends of a group is a quasi-isometry invariant, which captures the
large scale behavior at infinity of a group G. The ends are counted on graphs,
e.g. Cayley graphs, on which a group acts geometrically. The next Lemma and
Definition are taken from Meier's book [Mei08, Chapter 11.4].

Suppose Γ is a graph. For $n \in \mathbb{N}$ let in the following $B_n(v)$ be the ball of radius n around the vertex v in Γ. Let further $\|\Gamma \setminus B_n(v)\|$ denote the number of connected, unbounded components in the complement of $B_n(v)$. When talking about bounded, respectively unbounded, subsets in graphs we view graphs as metric spaces and consider the usual notion of (un-)bounded subsets of a metric space.

Lemma 7.39 *Let Γ be a connected, locally finite graph. Then, for all $m < n$, one has*

$$\|\Gamma \setminus B_m(v)\| \leq \|\Gamma \setminus B_n(v)\|.$$

Proof The balls $B_n(v)$ and $B_m(v)$ differ by a finite number of vertices and edges as Γ is locally finite. Let C be an unbounded, connected component of $\Gamma \setminus B_m(v)$. When $B_n(v)$ is removed either C remains unbounded and connected or C splits into several components. In the first case it still contributes one component to the count as removing a finite number of vertices does not get rid of the unbounded, connected component entirely. In the latter case, where C does not stay connected, it could contribute more than one component to the count. Hence, the lemma. $\square$

Definition 7.40 (Number of Ends of a Graph and a Group) Let Γ be a connected, locally finite graph and v a fixed vertex in Γ. The *number of ends* of Γ is given by

$$e(\Gamma) = \lim_{n \to \infty} \|\Gamma \setminus B_n(v)\|.$$

A finitely generated group G with finite generating set S has m *ends* if

$$e(G) := e(\Gamma(G, S)) = m.$$

The limit in the definition of the number of ends of a group exists by Lemma 7.39. It might, however, be equal to ∞. For more details and examples of ends of graphs see Chapter 11 in [Mei08]. The number of ends of a group is independent of the generating set. For a proof of this fact see [BH99, I.8.29] or [Mei08, Thm. 11.23].

Theorem 7.41 (Ends of a Group) *The number of ends $e(G)$ of a finitely generated group G is independent of the choice of a finite generating set S of G. Moreover, $e(G)$ is a quasi-isometry invariant.*

Example 7.42 Using Theorem 7.41 we compute the number of ends for some examples of groups by looking at Cayley graphs.

1. All finite groups have 0 ends as there are no infinite connected components of their Cayley graphs. A Cayley graph of Sym(3) is shown in Fig. 7.5. It is not hard to check that this graph has no ends.
2. The free abelian group $\mathbb{Z}^2$ on two generators has one end. Consider the Cayley graph of $\mathbb{Z}^2$ with respect to the standard generators as shown in Fig. 7.5c.

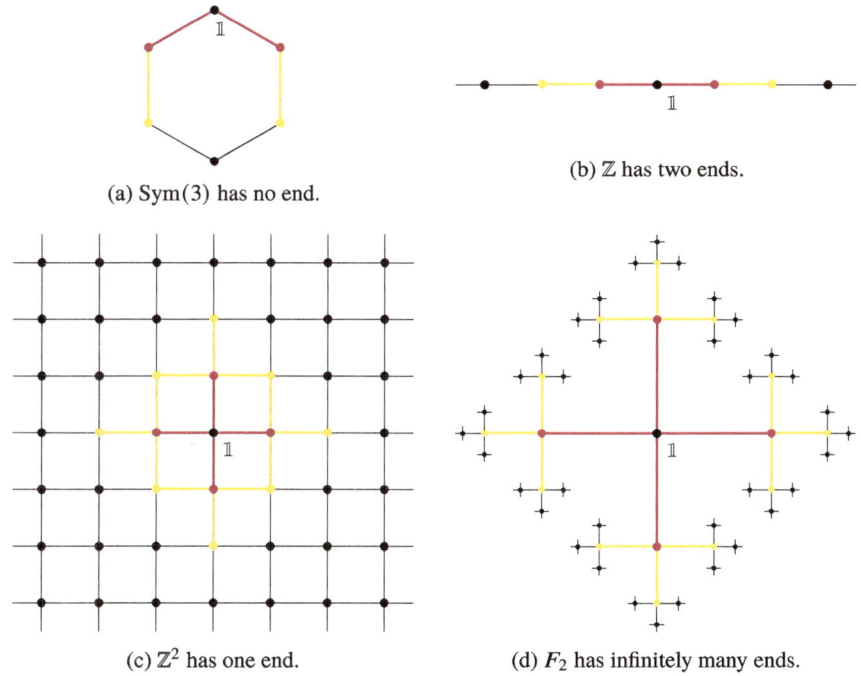

(a) Sym(3) has no end.

(b) $\mathbb{Z}$ has two ends.

(c) $\mathbb{Z}^2$ has one end.

(d) F_2 has infinitely many ends.

Fig. 7.5 Some Cayley graphs together with balls of radius one (red) and two (yellow) around the identity element, illustrating Example 7.42. (**a**) Sym(3) has no end. (**b**) $\mathbb{Z}$ has two ends. (**c**) $\mathbb{Z}^2$ has one end. (**d**) F_2 has infinitely many ends

Removing any ball of finite radius from the Cayley graph does not make the graph disconnected. Hence it has one end and so does $\mathbb{Z}^2$.

3. The group $G = \mathbb{Z}$ has two ends. Take the standard generating set consisting of a single element and the identity as a base vertex. Then removing balls of finite radius around the identity in the Cayley graph leaves you with two infinite connected components. Compare Fig. 7.5b.

4. The free group $F_2 = \langle a, b \rangle$ is an example of a group with infinitely many ends. Its Cayley graph with respect to the set $\{a, b\}$ is in Fig. 7.5d. Take the identity as base vertex and remove larger and larger balls around the origin. This leaves you with a growing number of connected components. Hence F_2 has infinitely many ends.

What about examples for groups with 3,4,5,... ends? It turns out there are none. Freudenthal [Fre45] and Hopf [Hop44] observed that the possibilities for the number of ends of a group are very restricted.

Example 7.42 shows that there exist groups with 0, 1, 2 or infinitely many ends. We will prove that there is no group G with $3 \leq e(G) < \infty$.

Theorem 7.43 (Freudenthal–Hopf Theorem) *The number of ends of any finitely generated group G satisfies*

$$e(G) \in \{0, 1, 2, \infty\}.$$

We follow the proof provided on pages 211 and 212 in [Mei08].

Proof Suppose for a contradiction that G is a group with a Cayley graph Γ with $k \geq 3$ ends. This in particular implies that G is an infinite group and that there is some $n \in \mathbb{N}$ and a vertex v in Γ such that $\Gamma \setminus B_n(v)$ has at least three unbounded connected components.

Recall from Example 2.33 that G acts by left-multiplication on any of its Cayley graphs. The action of G on the vertices of Γ is transitive. We may hence find an element $g \in G$ such that $g.v$ is contained in an unbounded connected component of $\Gamma \setminus B_n(v)$ and such that $d(v, g.v) > 2n$. But, this also implies that the n-balls around v and $g.v$ are disjoint.

The ball $g.B_n(v)$ is connected, since the action on the Cayley graph is simplicial. By the assumptions on $g.v$ we may conclude that the entire ball $g.B_n(v)$ is in an unbounded connected component C of $\Gamma \setminus B_n(v)$. But then, C is divided into at least k connected components, at least $(k-1)$ of which are connected. This follows again from the transitivity of the action.

Put $B := B_n(v) \cup g.B_n(v)$. Then B is a finite subgraph of Γ and $\Gamma \setminus B$ has at least $2k - 2$ connected components. But then, $\|(\Gamma \setminus B)\| \geq 2k - 2$. This implies that $e(G) = e(\Gamma) \geq 2k - 2 > k$, since $k \geq 3$. But, this contradicts the assumption that $e(G) = k$.   □

Hopf [Hop44] classified the groups having a given finite number of ends.

Theorem 7.44 (Recognizing Groups via Ends) *Let G be a finitely generated group. Then*

1. *G has no ends if and only if G is finite.*
2. *G has exactly two ends if and only if it is virtually $\mathbb{Z}$.*

The groups with two ends are virtually infinite cyclic and hence contain an infinite finite index subgroup generated by a single element. The groups with exactly one or infinitely many ends provide the richest classes.

Theorem 7.45 (Groups with Infinitely Many Ends [Sta68]) *For a finitely generated group G the following are equivalent.*

1. *$e(G) = \infty$*
2. *G is either isomorphic to an amalgamated free product $A \star_C B$ or to an HNN-extension $G = A_{\star C}$. In both cases C is finite and the indices satisfy $[A : C] \geq 3$ and $[B : C] \geq 2$.*
3. *G admits an action without global fixed points on a simplicial tree such that all edge-stabilizers are finite.*

The equivalence of the second and last item in Stallings result are shown using Basse-Serre theory. Note that Niblo [Nib04] gave a very nice geometric proof of Stallings theorem. Compare also Section 8.2 in Löh's book [Lö17].

We now turn to a relative notion of ends of a group.

Definition 7.46 (Ends Relative to a Subgroup) Let G be a finitely generated group and H a subgroup in G. The number of *ends of G relative to H*, denoted by $e(G, H)$, is the number of ends $e(\Gamma/H)$ of the quotient of the Cayley graph Γ of G by the natural left-action of H on Γ.

Equivalently we could define $e(G, H)$ to be the number of ends $e(X)$ where X is the graph with vertices the right-cosets of H with edges between Hx and Hxs for all $s \in S$, where S is a finite generating set of G.

The following theorem by Sageev [Sag95] on the number of relative ends with respect to hyperplane stabilizers in groups acting on CAT(0) cube complexes is crucial for the proof of the Tits alternative.

Theorem 7.47 (Relative Ends of Groups Acting on Cube Complexes) *Suppose G is a group acting without a global fixed point on a* CAT(0) *cube complex X of finite dimension. Then there exists a hyperplane J in X such that $e(G, \mathrm{Stab}_G(J)) > 1$.*

In addition to Stallings' theorem we have the following result, which is due to Dunwoody and Swenson [DS00].

Theorem 7.48 (Algebraic Torus Theorem) *Let G be a finitely generated group and $H < G$ a subgroup. If H is a virtually polycyclic group with $e(G, H) > 1$ then one of the following possibilities holds*

1. *G is virtually polycyclic.*
2. *There exists a short exact sequence $\mathbb{1} \to P \to G \to G/P \to \mathbb{1}$ where P is virtually polycyclic and G/P a non-elementary Fuchsian group.*
3. *G decomposes as a nontrivial amalgamation $A \star_C B$ or HNN-extension $A_{\star C}$ over a virtually polycyclic group C.*

Let me introduce a notion mentioned in this theorem, which plays a crucial role in the proof of the Tits alternative.

Definition 7.49 (Polycyclic Groups) A group G is *polycyclic* if there exists a sequence of subgroups $G = G_0, G_1, \ldots, G_n = \mathbb{1}$ such that G_{i+1} is normal in G_i for all i and the quotient G_i / G_{i+1} is cyclic for all i.

A good way of thinking about a polycyclic group is to view it as a "tower of cyclic groups". This point of view is justified by the fact that iterated semidirect products of cyclic groups are polycyclic. Other examples of polycyclic groups include finitely generated abelian or nilpotent groups.

Every polycyclic group is solvable and satisfies in addition that all its subgroups are finitely generated.

7.2.3   Fuchsian Groups

Discrete actions on the hyperbolic plane $\mathbb{H}^2$ play a crucial role in what follows. One natural class of groups that admits such actions is the class of Fuchsian groups, which we now introduce.

Definition 7.50 (Fuchsian Groups) A group G is *Fuchsian* if G is a discrete subgroup of $PSL_2(\mathbb{R})$.

The group $PSL_2(\mathbb{R})$ acts on $\mathbb{H}^2$ hence so does any Fuchsian group. Consider the action of a Fuchsian group on the Poincaré disc model D of the hyperbolic plane D. Viewing D as a disk in the Euclidean plane the orbit $G.z$, for some $z \in D$, under the G–action may have limit points in the boundary circle. One can prove the following property.

Proposition 7.51 *The set of limit points of the action of G on D is independent of the choice of the basepoint z. The group G either has $0, 1, 2$ or ∞ limit points. Moreover, the number of limit points equals the number of ends of G.*

This proposition allows us to talk about limit points of a group.

Example 7.52 The group $PSL_2(\mathbb{Z})$ is a Fuchsian group having ∞-many limit points. These points correspond to the ends of the Cayley graph, when embedded as a dual graph of the induced tiling in $\mathbb{H}^2$. Compare Fig. 6.10 and the dual graph inside the tiling of the upper half-plane.

Definition 7.53 (Elementary Fuchsian Group) We say that G is *elementary Fuchsian* if it has a finite number of limit points.

In case such a Fuchsian group G has infinitely many limit points it contains a copy of the free group F_2 on two generators. One can introduce a notion of an end via equivalence classes of certain rays in the Cayley graphs. The Cayley graph of F_2 may then be embedded in D in such a way that it's ends correspond to the limit points.

Lemma 7.54 *Every non-elementary Fuchsian group virtually contains a copy of the free group on two generators F_2.*

We make use of this fact in the proof of the Tits alternative.

7.2.4   The Tits Alternative

We have collected all the relevant material to prove the Tits alternative for groups acting nicely on CAT(0) cube complexes.

Theorem 7.55 (Tits Alternative for Cube Complexes) *Let G be a finitely generated group acting properly discontinuously on a nonempty, finite dimensional*

CAT(0) *cube complex X and suppose that there is a bound on the order of its finite subgroups. Then every subgroup $H < G$ is either virtually solvable or contains a (non-abelian) free subgroup of rank 2. That is, G satisfies the Tits alternative.*

Remark 7.56 The assumption that X is finite dimensional can, in general, not be dropped: Thompson's group F acts on an infinite dimensional CAT(0) cube complex, but it is known not to satisfy the Tits alternative. Similarly one can, in general, not drop the assumption on the bound of the order of subgroups of G. To see this consider the following example: Let $G = G_1 \subset G_2 \subset \ldots$ be an ascending sequence of finite groups G_i. One can construct a coset tree T associated with G from the left-cosets of the subgroups G_i in G. See [Ser03] for details. By construction G admits a proper left-action on this tree T. However, G does not satisfy the assertion of Theorem 7.55.

We need two additional statements about semisimple actions.

Definition 7.57 (Semisimple Action) A *semisimple action* is an action for which each group element is a *semisimple isometry*. That is, there exists $x_0 \in X$ such that $d(g.x_0, x_0)$ equals the translation length $|g| := \inf\{d(g.x, x) \mid x \in X\}$ of g.

The class of semisimple isometries of a metric space is split into two subclasses. Isometries f with $|f| > 0$ are called *hyperbolic* and isometries with translation length $|f| = 0$ are *elliptic*. Non-semisimple isometries with translation length equal to zero are called *parabolic*. Note that there can also be non-semisimple isometries with positive translation length.

Lemma 7.58 (Bridson's Lemma)

1. *Suppose G admits a cellular action on a* CAT(0) *complex X, which is built out of finitely many shapes. Then the action of G on X is semisimple with a discrete set of translation lengths.*
2. *Suppose G admits a semisimple, properly discontinuous action on a* CAT(0) *space X. Then every virtually solvable subgroup of G is virtually abelian. (In particular G is virtually abelian if it is virtually solvable.)*

In addition we have the following theorem. See Theorem 7.5 and Remark 7.7 in [BH99] or Theorem 2.4 in [SW05]. The group G appearing in the statement is not necessarily finitely generated. And indeed, the non-finitely generated case of the proof of Theorem 7.61 is the only place where we use this result.

By a *flat of dimension n* in a metric space X we mean the image of an isometric embedding of the n-dimensional euclidean space.

Theorem 7.59 (Properties of Semi-simple Actions) *Let G act by semi-simple isometries on a* CAT(0) *space such that*

1. *there is a bound on the dimension of an isometrically embedded flat,*
2. *the set of translation numbers of elements of G is discrete at* 0,
3. *there is a bound on the order of finite subgroups.*

Then any sequence $H_1 \subsetneq H_2 \subsetneq \cdots$ of virtually abelian subgroups terminates.

By setting H_i to $H^{t^{-i}}$ we obtain:

Corollary 7.60 *Let G act as in Theorem 7.59. If $H^t \subset H$ then $H^t = H$.*

The following theorem directly implies the Tits alternative as stated in Theorem 7.55. To see this apply Theorem 7.61 to any subgroup H of G in the setting of Theorem 7.55.

Theorem 7.61 (Abelian Versus Free) *Let G be a (not necessarily finitely generated) group acting properly discontinuous on a finite dimensional cube complex X and suppose G has an upper bound on the order of its finite subgroups. Then either*

1. *G contains a free subgroup of rank two, or*
2. *G is virtually finitely generated abelian.*

Note that finitely generated abelian groups are polycyclic and hence solvable.

Proof Suppose first that G is not finitely generated. Thus G contains an infinite ascending sequence of subgroups $G_1 \subsetneq G_2 \subsetneq \ldots$. In case every G_i contains a free subgroup of rank 2 we are done and case 1 holds.

To finish the proof in the non-finitely generated case assume for a contradiction that each G_i is virtually finite abelian. We verify conditions 1. to 3. in Theorem 7.59. The first item follows from the fact that the cube complex X is itself finite dimensional. Observe that a finite dimensional cube complex is a polyhedral complex with finitely many shapes. Then item 1 of Lemma 7.58 implies item 2 of Theorem 7.59. The last item is put as an assumption on G. We may thus apply Theorem 7.59 to obtain that the given sequence of subgroups terminates, which is impossible.

The proof for G finitely generated is by induction on $n = \dim(X)$. For $n = 0$ the group G has to be finite since otherwise it can not act properly discontinuously on a set of points. We are hence in Case 2. Suppose the theorem is true for all $k = \dim(X) < n$. Then again, if G is finite we are in Case 2. Hence, without loss of generality we may assume that G is infinite. By assumption G does not have a global fixed point on X and we may apply Theorem 7.47.

Put $H := \mathrm{Stab}_G(J)$ with J the hyperplane obtained from the theorem just mentioned. The action of H on the CAT(0) cube complex J is properly discontinuous and the other assumptions of the induction hypothesis are satisfied. Since we have already proven the statement for non-finitely generated groups and cube complexes of smaller dimensions we may deduce that H either contains a free subgroup of rank two or is virtually finitely generated abelian.

Suppose Case 1 applies. Then there exists $F_2 < H < G$ and the assertion follows. If Case 2 is satisfied by H, then H is in particular virtually polycyclic and we may apply the algebraic torus Theorem 7.48. Hence, G satisfies one of the following three possibilities:

a) G is virtually polycyclic,
b) G contains a non-elementary abelian Fuchsian quotient with virtually polycyclic kernel, or

c) G decomposes as a non-trivial product (free amalgamation or HNN-extension) over a virtually polycyclic subgroup.

We are dealing with each of the three possibilities separately and show that in any case G satisfies the assertion of our theorem.

Case a): We apply Bridson's Lemma 7.58 to prove that G is virtually abelian. Item 1 of Lemma 7.58 implies that Lemma 7.58 Item 2 is applicable. Since being (virtually) polycyclic implies being (virtually) solvable the assertion follows.

Case b): This is the easiest case as the non-elementary Fuchsian quotient does contain a copy of F_2. Hence, G does by Lemma 7.54.

Case c): Suppose $G = A \star_P B$ or $G = A\star_P$ for some virtually polycyclic P. Then by Item 1 of Lemma 7.58 the action of G on X is semisimple. Item 2 of Lemma 7.58 thus implies that P needs to be virtually abelian.

Let us first consider the case with $G = A \star_P B$. The normal form theorem, stated as Theorem 7.37, implies that if $[A : P] > 2$ or $[B : P] > 2$ then G does contain a free subgroup of rank two. We may hence suppose that $[A : P] = 2 = [B : P]$. In this case the Bass-Serre tree T is a line and any edge in T is stabilized by a conjugate of P. This implies then that there exists an index two subgroup G' of G, which acts by translation on T such that $\ker(G' \to \mathbb{Z}) \cong P$. Hence, $G' \cong P \rtimes \mathbb{Z}$ is polycyclic and G is thus virtually polycyclic and (by Lemma 7.58) virtually abelian.

Suppose now $G = C\star_P$. Then there exist subgroups P_1, P_2 of C isomorphic to P via $\phi : P_1 \to P_2$ isomorphic to one another. Remark 7.38 after the normal form Theorem 7.35 now implies that if $[C : P_1] > 1$ and $[C : P_2] > 1$ then G contains a free subgroup of rank two. One can show that if $[C : P_1] = 1$ then $[C : P_2] = 1$ and vice versa. And we may conclude that G is isomorphic to $P \rtimes \mathbb{Z}$, is virtually polycyclic an hence virtually abelian. □

7.3 Special Cube Complexes

Special cube complexes saw much attention when in 2012 Ian Agol [Ago13] proved the virtual Haken and virtual fibration conjecture, deep open problems in 3-manifold theory, using CAT(0) cube complexes. Agol's work builds up on work of Dani Wise [Wis02] and many others, for example, [HW08] and [AGM09].

Wise's *cubical route to understanding groups* [Wis14] provides a strategy to understand groups by making them act on special cube complexes. First one shows that a group G acts geometrically on a CAT(0) cube complex X. This is often done by explicit constructions of such spaces. Then one needs to prove that the quotient cube complex X/G is (virtually) special. By work of Haglund and Wise [HW08] this has then may consequences for the group.

This section is loosely based on lecture notes of Wilton [Wil17] and the work of Haglund and Wise [HW08]. We introduce special cube complexes and get to know some of their properties. We define special cube complexes in the first subsection and prove some elementary properties. The second subsection contains a proof of the fact that every special cubulated group embeds in a RAAG. The last subsection provides more detailed comments on Agol's result.

7.3.1   What Makes a Cube Complex Special?

In this subsection special cube complexes are introduced.

We say that two hyperplanes H and H' *intersect* if they intersect as subsets of the ambient (metrized) cube complex X. Alternatively one can define $H \cap H' \neq \emptyset$ to hold if and only if the natural embedding of $H \sqcup H'$ into X is not injective.

Definition 7.62 (Properties of Hyperplanes) Let H and H' be two distinct hyperplanes in a (metrized) cube complex X. Then

1. H is *two-sided* if the support $N(H)$ is a product of the form $I \times H$ with I an interval.
2. H *self-intersects* if there exist two distinct mid-cubes M_1 and M_2 in H such that the intersection $M_1^\circ \cap M_2^\circ$ of their relative interiors is nonempty. Or, equivalently, if the natural inclusion map $H \to X$ is not injective.
3. H *self-osculates* if it is two-sided and there exist two edges in X both perpendicular to H that share a vertex and are not contained in a common cube.
   We say that H *directly* self-osculates in case the other two vertices of these edges are on a same side of H. Moreover H *indirectly* self-osculates if it does not directly self-osculate.
4. H and H' *inter-osculate* if they intersect and there exist two distinct edges e, e' in X with the following properties: e is perpendicular to H and e' to H', both edges share a vertex but are not contained in a common cube.

Two examples of self-intersecting and of two-sided, directly self-osculating hyperplanes are shown in Fig. 7.6.

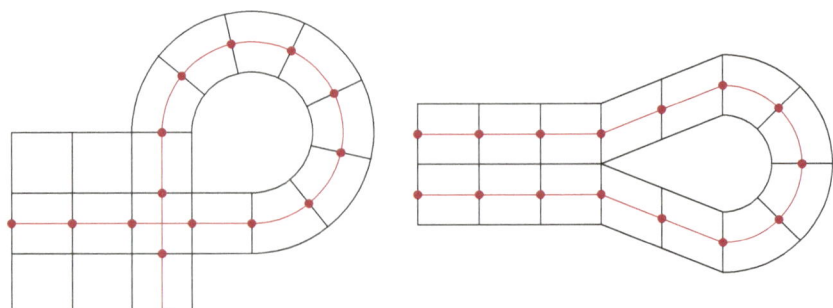

Fig. 7.6 On the left a self-intersecting and on the right a two-sided, directly self-osculating hyperplane

Cube complexes with particularly well behaved hyperplanes have a big impact on groups acting on them and hence deserve their own name.

Definition 7.63 (Special Cube Complexes) A cube complex is *special* if all of the following are satisfied:

1. every hyperplane is two-sided,
2. no hyperplane self-intersects,
3. no hyperplane directly self-osculates and
4. no two hyperplanes inter-osculate.

Each of the four conditions in the definition of a special cube complex lifts to the hyperplanes in the universal cover. Hence if X is special any cover of X is also special. Compare Exercise 7.101.

Remark 7.64 Names vary a bit from source to source. What we call a special cube complex is called an *A-special* cube complex by Haglund and Wise in [HW08]. In their notion of a special cube complex hyperplanes are not assumed to be two-sided. Every special cube complex in their sense, which has a finite number of hyperplanes (e.g. cocompact ones) have an A-special cover. Hence this is not an important distinction. A similar definition to the one introduced in this book was also used in [Wil17] and [Sag14]. However, Wilton also assumes a special cube complex to be locally non-positively curved. Note that Haglund and Wise [HW08] in addition assume special cube complexes to be simple which means that links are simplicial complexes. We have this as a standing assumption on every cube complex.

Any graph is (trivially) special. More interesting examples of special cube complexes are provided by the Salvetti complexes of RAAGs. As Salvetti complexes themselves are not even cube complexes (edges and cubes have self-gluings) we need to cubically baricentrically subdivide them in order to obtain cube complexes. Those however, turn out to be special.

Proposition 7.65 *Let Γ be a graph defining a right-angled Artin group $A(\Gamma)$. The cubical barycentric subdivision $\hat{S}(\Gamma)$ of the Salvetti complex of $A(\Gamma)$ is a special cube complex, as is its universal cover.*

Proof One can easily check that the barycentric subdivision $\hat{S}(\Gamma)$ is a cube complex. We prove that it is special. In order to do so we extend the coloring of the edges of $S(\Gamma)$ to its barycentric subdivision as follows. Every edge in the Salvetti complex $S(\Gamma)$ is colored by a generator. The barycentric subdivision has more edges. Some of them are subdivisions of edges in $S(\Gamma)$, others appear as new edges inside a square (or higher dimensional cube) connecting midpoints of faces to the midpoint of the cube. For the subdivided edges we keep the original colors for both new edges in the barycentric subdivision. Every new edge, say in a square, is parallel to an edge that was already present in the original complex $S(\Gamma)$. Parallel edges (inside every square) of $S(\Gamma)$ have the same color. We preserve this property and color the new edges so that in $\hat{S}(\Gamma)$ this property is still satisfied.

We now verify the four conditions of Definition 7.63. Parallelism is preserved by the gluing rules of $S(\Gamma)$ and hence also holds in the cube complex $\hat{S}(\Gamma)$. Thus, all hyperplanes are two-sided. To see the second condition we argue by contradiction. So suppose that there exists a self-intersecting hyperplane in $\hat{S}(\Gamma)$. Then there exists a square M where all the edges are labeled with the same generator of the RAAG A_Γ. Such a square corresponds to a relation of the form $s^4 = \mathbb{1}$. As such a relation will never occur in the presentation of a RAAG we arrive at a contradiction and no hyperplane self-intersects.

Suppose now a hyperplane H in $\hat{S}(\Gamma)$ self-osculates. Then there exist two distinct edges e, e' both perpendicular to H, which share a vertex. But then, e and e' can not have the same label as edges sharing a vertex have different labels by construction. As every hyperplane corresponds to a unique label in a Salvetti complex this yields a contradiction.

To see the last item suppose that there exist two distinct inter-osculating hyperplanes H_a and H_b, defined by edges e_a and e_b labeled a and b, respectively. Then e_a and e_b bound a square at the place where H_a and H_b intersect in a same cube. But, at the same time, there exists a pair of edges e'_a and e'_b also labeled a, respectively b, that share a vertex but do not bound a square. This is impossible given how Salvetti complexes are constructed. Hence $\hat{S}(\Gamma)$ is special.

The fact that also the universal cover is special is easy to check. For a reference see [HW08, Cor. 3.8].   □

We follow [HW08, Lemma 3.5] and provide a method to construct new special cube complexes out of old.

Lemma 7.66 *Let X_1 and X_2 be cube complexes. If both satisfy one of the following properties, then $X_1 \times X_2$ also has that property:*

1. *Each hyperplane is embedded.*
2. *Each hyperplane is two-sided.*
3. *Each hyperplane embeds and no hyperplane directly self-osculates.*
4. *Each hyperplane embeds and no hyperplane self-osculates.*
5. *No two hyperplanes inter-osculate.*

Proof Without loss of generality we may restrict to connected complexes X_i. We start by describing (oriented) walls in $X_1 \times X_2$. Recall from Definition 5.1 that the wall $W(e)$ corresponding to an edge e in a cube complex is the set of all edges in X perpendicular to the hyperplane $H(e)$ defined by e. We may now orient each edge in a cube complex by declaring one of its vertices as the source and the other the sink. We say that an edge is oriented from the source to the sink.

Observe first that an (oriented) edge in $X_1 \times X_2$ is of the form $\vec{a}_1 \times v_2$ or $v_1 \times \vec{a}_2$ where $\vec{a}_i$ is an (oriented) edge in X_i and v_j a vertex in X_j. Now the walls in $X_1 \times X_2$ have a similar structure. The wall through $\vec{a}_1 \times v_2$ is of the form $W(\vec{a}_1) \times X_2^0$, where X_2^0 denotes the set of vertices in X_2.

Suppose that such a wall in $X_1 \times X_2$ self-intersects. Then there are edges a_1, b_1, which are parallel in X_1 and vertices v_2, w_2 in X_2 such that the edges $a_1 \times v_1$ and

$b_1 \times w_1$ are consecutive (hence share a vertex) in some square in the product. But then, $v_2 = w_2$ and a_1 and b_1 must be consecutive in a square of X_1 and X_1 contains a self-intersecting hyperplane.

To see the second item suppose that the hyperplane corresponding to the wall $W(a_1) \times X_2^0$ is one-sided. Fix a vertex v_2 in X_2. One-sidedness implies that the oriented edge $\vec{a}_1 \times v$ is parallel to the edge $\bar{a}_1 \times v_2$, where the edge a_1 is oriented in reverse. Choose a sequence of elementary parallel edges inside squares in $X_1 \times X_2$ that connects $\vec{a}_1 \times v$ with $\vec{a}_1 \times v_2$. Forgetting the second factor we obtain a sequence of elementary parallel edges inside squares of X_1 yielding $\vec{a}_1$ is parallel to $\vec{a}_1$. Hence X_1 is one-sided.

To see the third item suppose that the hyperplane associated with the wall $W(a_1) \times X_2^0$ directly self-osculates. There exist then a sequence of elementary parallelisms between oriented edges $\vec{a}_1 \times v_2$ and $\vec{b}_1 \times v_2$ such that $\vec{a}_1$ and $\vec{b}_1$ start in the same vertex v_1, such that a_1 and b_1 are parallel to a common edge $\vec{a}$ and such that there is no square in $X_1 \times X_2$ in which $\vec{a}_1 \times v_2$ and $\vec{b}_1 \times v_2$ are consecutive. As for the second item project a sequence of elementary parallelisms in $X_1 \times X_2$ to the first component to obtain a sequence of elementary parallelisms between $\vec{a}_1$ and $\vec{b}_1$. But then, either $W(\vec{a}_1)$ self-intersects or directly self-osculates in X_1.

The same arguments imply the fourth item.

To finally prove the last item suppose that W and V are two non-equal walls in $X_1 \times X_2$. Then either W and V are both walls defined by products of the form an edge times a vertex where the edges are in the same cube complex X_i or, alternatively, one in X_1 and the other in X_2. Consider first the case, where the edges are in the same X_i, say for $i = 1$. That is, $W = W(a_1 \times v_2)$ and $V = V(b_1 \times w_2)$ where a_1, b_1 are edges in X_1 and v_2, w_2 vertices in X_2. Then a square in which W and V has to be of the form $C \times u_2$ with C a square in X_1 and u_2 a vertex in X_2. As otherwise, if the intersecting square is of the form edge times edge the walls V and W would have to agree. But then, the walls $W(a_1)$ and $W(b_1)$ intersect in X_1 in the square C. In the second case $W = W(a_1 \times v_2)$ and $V = V(v_1 \times a_2)$. We claim that such a pair of walls never osculates in the product $X_1 \times X_2$. If w_i is a vertex of a edge b_i, the product edges $b_1 \times w_2$ and $w_1 \times b_2$ belong to the square $b_1 \times b_2$. Hence the claim and thus the last item in the lemma.   □

From Lemma 7.66 one can deduce that products of special cube complexes are also special. See Exercise 7.100.

Every special cube complex can be completed to a cube complex with local non-positive curvature. We state this fact as Proposition 7.73 and introduce a few more technical terms needed for the proof.

Definition 7.67 (Horizontal Edge-Paths in Strips) Write I_n for a graph consisting of n consecutive edges and write B_n for the strip $I_n \times I_1$ consisting of n consecutive squares. We call the subgraph $I_n \times \{0\}$ the *bottom horizontal path* of B_n and $I_n \times \{1\}$ the *top horizontal path* of B_n.

Definition 7.68 (Locally Special Cube Complex) A cube complex X is called *locally special* if the following two conditions are satisfied.

1. A combinatorial map that sends the bottom horizontal path of B_2 to two consecutive edges of a square in X also sends the top horizontal path to two consecutive edges of a square.
2. Any combinatorial map $c : B_4 \to X$ that sends the bottom horizontal path to the boundary of a square in X identifies the two edges $\{0\} \times [0, 1]$ and $\{1\} \times [0, 1]$.

The subsequent Lemmas are taken from [HW08]. We start with the following preparatory lemma, which is a direct consequence of Definitions 7.63 and 7.68.

Lemma 7.69 *Suppose X is a cube complex such that no hyperplane directly self-osculates and no pair of hyperplanes inter-osculate. Then X is locally special. In particular every special cube complex is locally special.*

Proof The second item in the definition of locally special cube complexes implies that any non-locally special cube complex has a pair of inter-osculating hyperplanes.

Let c be a combinatorial map that sends the bottom horizontal path of B_2 to two consecutive edges of a square in X. If c does not send the top horizontal path to two consecutive edges of a square it can happen that the image is part of a directly self-osculating and also self-intersecting hyperplane. This hyperplane is then defined by the images of the two edges in the bottom horizontal path. So if X is not locally special it can have a self-intersecting and directly self-osculating hyperplane. □

The following Lemma is a simple consequence of the definition of locally special complexes. We leave the proof to the reader, see Exercise 7.103.

Recall the notion of a k-corner from Definition 4.48 and use notation as in Notation 4.47.

Lemma 7.70 *Let X be a cube complex. Then X is locally special if and only if the following two conditions are satisfied for all $3 \leq k \in \mathbb{N}$:*

1. *For every edge e containing a vertex v_k every k-corner containing e has a k-corner adjacent along e. This k-corner is unique and equals $\sigma_e(c)$.*
2. *For every pair of edges e, e' containing v_k and every k-corner $c : C_k \to X$ we have $\sigma_e(\sigma_{e'}(c)) = \sigma_{e'}(\sigma_e(c))$.*

Please note that the next lemma assumes the cube complex to be locally special, but it does not make any restrictions on the size k of the cubes. For its proof we need the following notation.

Notation 7.71 Using Notation 4.47 define groups G_k for $k \in \mathbb{N}$ as follows: let G_k be the group generated by the σ_e for all edges e in X containing the vertex v_k. As relations take $\sigma_e^2 = \mathbb{1}$ and $\sigma_a \sigma_b = \sigma_b \sigma_a$ for all edges a, b. For any $v \in I^k$ denote by σ_v the unique element in G_k that maps v_k to v.

Lemma 7.72 *Every k-corner $D_k \to X$ in a locally special cube complex X extends uniquely to a combinatorial map $C^{k,2} \to X$ from the 2-skeleton of a k-cube into X.*

Proof Let $d : D_k \to X$ be a k-corner. The group G_k acts on the set of k-corners and $\sigma_e(c) = c \circ \sigma_e$ on the union of squares containing e. Consider for each vertex $v \in I^k$

the action of G_k to define a new corner $c_v = \sigma_v(c)$. Now define a new combinatorial map on the whole k-cube by putting $q^v = c_v \circ \sigma_v^{-1}$.

If v, w are adjacent vertices let e denote the edge of T_k such that $\sigma_w = \sigma_v \sigma_e$. Then we have $q^w = c_w \circ \sigma_e^{-1} \circ \sigma_v^{-1}$. But, c_v and c_w are adjacent k-corners along the edge e and hence $q^v = q^w$ on the union of the squares containing both vertices v and w. This implies that the maps q^v induce a combinatorial map $q : C^k \to X$ extending c.

As links in a cube complex are simplicial complexes any two k-corners with the same vertex adjacent along the same edge are equal by Remark 4.49. This implies that the extension is unique. $\square$

Recall from Definition 4.46 the definition of a completable cube complex.

Proposition 7.73 *Let X be a special cube complex. Then X is completable and hence contained in a unique, smallest locally* CAT(0) *cube complex with the same 2-skeleton as X.*

Proof As a special cube complex X is locally special by Lemma 7.69. We show that locally special cube complexes are completable. Apply Lemma 4.55 to X and obtain a complex $\bar{X}$ together with a combinatorial map $j : X \to \bar{X}$ that restricts to an isomorphism of the 2-skeleta. By construction every combinatorial map $C^{k,2} \to \bar{X}$ extends to a map on the k-cube to X. So there are no empty k-cubes in $\bar{X}$ for $k \geq 3$. Let now $c : D_k \to \bar{X}$ be a k-corner of $\bar{X}$. Then, using the fact that X is locally special, Lemma 7.72 implies that there exists a unique extension of c to the 2-skeleton of a k-cube, which then extends to a unique k-cube. Hence, there are no empty k-cubes in $\bar{X}$ for $k \geq 2$ and X is locally CAT(0). $\square$

7.3.2   Special Cubulated Groups Embed in RAAGs

Salvetti complexes of RAAGs were our first examples of spaces giving rise to special cube complexes in nature. In Proposition 7.65 we showed that their cubical barycentric subdivision is a special cube complex. We now aim to show, compare Theorem 7.79, that in fact every compact special cube complex embeds into a (finite cover of) a barycentric subdivision of a Salvetti complex. This is done using hyperplane graphs, which we define now.

Definition 7.74 (Hyperplane Graph) Let X be any cube complex. Define its *hyperplane graph* $\Gamma(X)$ to be the graph with vertex set the set of all hyperplanes in X. Put an edge between a pair of distinct hyperplanes H_1 and H_2 if and only if H_1 and H_2 intersect.

In case X has finitely many hyperplanes we may take the hyperplane graph $\Gamma(X)$ as input for the definition of a RAAG and consider the associated Salvetti complex $S(\Gamma(X))$. It turns out that the cube complex locally embeds into the Salvetti complex.

Definition 7.75 (Characteristic Map) Let X be a cube complex with finitely many hyperplanes all of which are two-sided. Let $\Gamma = \Gamma(X)$ be the hyperplane graph of X and let $S = S(\Gamma)$ be the Salvetti complex of the RAAG defined by the hyperplane graph. Orient the edges in S and choose for every hyperplane H in X a consistent orientation of the edges in its wall, that is the edges perpendicular to H in X. The *characteristic map* $\varphi : X \to S$ is then uniquely determined by the following properties:

- φ maps the 0-skeleton of X to the unique vertex in S,
- the image $\varphi(e)$ of an edge e in X is obtained as follows: the edge e is mapped in an orientation preserving way to the unique edge in S which corresponds to the vertex $H(e)$ in Γ. Here $H(e)$ is the hyperplane in X defined by e.
- A higher-dimensional cube C in X corresponds to a collection H_i, $i = 1, \ldots, n$ of pairwise intersecting hyperplanes. Denote the corresponding edges in S by $e_1, e_2, \ldots, e_n$, respectively, and map C to the unique cube spanned by $\varphi(e_1), \ldots, \varphi(e_n)$ in $S(\Gamma(X))$ such that orientations of edges are preserved.

The characteristic map is well defined as we chose a consistent orientation on the collection of edges in every wall.

Remark 7.76 Note that the assumption that X has finitely many hyperplanes is not crucial for the definition of hyperplane graphs and characteristic maps. So far we only need to assume this, since we defined right-angled Artin groups with respect to finite graphs. All of these definitions easily extend to the infinite case. A compact cube complex for example satisfies the assumption of having finitely many hyperplanes.

Let us consider an example of such a cube complex and its characteristic map.

Example 7.77 Construct a cubical 2-torus $X = T^2$ out of the barycentric subdivision of a square as follows: cubically barycentrically subdividing a square yields four squares centered around a unique mid-vertex. Opposite sides of the original square are identified (without twisting) to form a 2-torus X. A similar construction was carried out in Example 4.29, see also Fig. 4.7 item (2), where we constructed a torus from a square subdivided into nine smaller squares.

The complex X has four vertices, eight edges and four squares. Moreover, X contains four hyperplanes and the corresponding hyperplane graph Γ is a 4-gon. The Salvetti complex $S = S(\Gamma)$ is a rose with four petals, one for each hyperplane in X, onto which a collection of four 2-tori is glued. It is not hard to check that each of the four vertices of X is contained in four edges corresponding to pairwise different hyperplanes.

The characteristic map sends each square in X to one of the 2-tori in the Salvetti complex S. One obtains that X is a covering space of S. Fibers of edges in S contain two elements, the fiber of the unique vertex in S contains all four vertices of X.

Proposition 7.78 *Let X be a complete special cube complex with finitely many hyperplanes. Then the characteristic map φ is a local isometric embedding.*

Proof Proposition 7.73 implies that X is completable and contained in a locally CAT(0) cube complex with the same 2-skeleton X. Since X is complete it is itself locally CAT(0). This implies that every point $p \in X$ has a neighborhood, which is isometric to a cone over its link $lk_X(p)$. Therefore it is enough to prove that φ induces an injective map $\tilde{\varphi}$ from links $lk_X(x)$ of vertices x in X to the link $lk_S(x_0)$ of the unique vertex x_0 in the Salvetti complex $S = S(\Gamma(X))$. This then implies the assertion.

Recall that we have the following correspondences: A vertex in $lk_X(x)$ corresponds to an edge in X, which in turn defines a hyperplane and hence a vertex in $\Gamma(X)$ and an edge in $S(\Gamma(X))$. Moreover, an edge in $lk_X(x)$ corresponds to a 2-cube in X, which yields a pair of intersecting hyperplanes in X and a cube in $S(\Gamma(X))$.

The fact that there is no self-osculating hyperplane in X implies that no two vertices at distance $\geq \frac{\pi}{2}$ in $lk_X(x)$ are identified under $\tilde{\varphi}$. That is, the hyperplanes defined by two distinct edges in X sharing a vertex are never equal.

Non-existence of inter-osculating hyperplanes implies that there do not exist vertices u, v in $lk_X(x)$ such that $d(u, v) > \frac{\pi}{2}$ and $d(\varphi(u), \varphi(v)) = \frac{\pi}{2}$.

Finally, by induction on the length of a shortest non-injective path, one can show that φ has to be injective on $lk_X(x)$. $\square$

This proposition allows to prove the following result by Haglund and Wise, compare Theorem 1.1 in [HW08].

Theorem 7.79 (Special Cube Complexes Locally Embed in Salvetti Complexes)
Let X be a compact, complete, non-positively curved cube complex. Then X is special if and only if there exists a graph Γ and a local isometric embedding of X into the Salvetti complex $S(\Gamma)$.

Proof Suppose first that X is special. Let Γ be the hyperplane graph. The characteristic map φ defined in Definition 7.75 is then, by Proposition 7.78, the desired locally isometric embedding.

To see the converse suppose that $f : X \to S$ is a local isometric embedding for some Salvetti complex $S = S(\Gamma)$. We prove by contradiction that X is special. If there exists a hyperplane H in X that self-intersects in a cube C this cube is mapped onto a cube of the same dimension in S by f. Hence the image $f(H)$ also self-intersects in S and induces a self-intersecting hyperplane in the cubical barycentric subdivision of the Salvetti complex. This contradicts Proposition 7.65. Using similar arguments for all other properties listed in Definition 7.63 the assertion follows. $\square$

It is left to the reader to prove that (universal) covers of special cube complexes are special, see Exercise 7.101. Combining this with the theorem above one obtains the following statement. For a proof see Corollary 5.9 of [Wil17].

Corollary 7.80 *The characteristic map lifts to an isometric embedding of the universal cover of X to the universal cover of the Salvetti complex, that is,*

$$\bar{\varphi} : \bar{X} \to \overline{S(\Gamma(X))}.$$

Moreover φ induces an injective homomorphism $\varphi_* : \pi_1(X) \to A_{\Gamma(X)} = \pi_1(S(\Gamma))$.

Finally, this implies (using Exercise 7.95) that subgroups of RAAGs are exactly the fundamental groups of special cube complexes.

Corollary 7.81 *A group G is a subgroup of a RAAG if and only if G is the fundamental group of a special cube complex.*

7.3.3 Cube Complexes Meet 3-Dimensional Manifolds

The theory of special cube complexes had a huge impact on low-dimensional topology. Work of Agol, Wise, and many others lead to a solution of two longstanding open conjectures: the virtual fibering and virtual Haken conjecture.

Theorem 7.79 provides a characterization of compact, special cube complexes. As a consequence special groups are identified as those that embed into a RAAG. But, it is still desirable to get a better understanding of this class. For example, does every hyperbolic group act on a special cube complex?

To understand group actions on cube complexes it is often useful to examine the stabilizers of hyperplanes. We have already seen in Sects. 5.1 and 6.3 that hyperplanes are an important tool when it comes to understanding the structure of a CAT(0) cube complex and to cubulate groups, that is, to get them act nicely (e.g. geometrically) on a CAT(0) cube complex.

Hyperplanes also turn out to be useful when one wants to characterize which groups admit an action on a special cube complex.

Haglund and Wise proved that a hyperbolic cubulated group is virtually special if and only if every hyperplane stabilizer is separable [HW08].

Agol proved that every hyperbolic cubulated group is virtually special. More precisely the following theorem holds:

Theorem 7.82 (Agol's Theorem) *Suppose G is a hyperbolic group that acts cocompactly and properly on a CAT(0) cube complex X. Then G has a finite index subgroup H such that the action of H on X is free and the quotient X / H is special.*

How does this relate to 3-manifold theory?

It is widely known that fundamental groups of closed 3-manifolds are hyperbolic. If one can make them act on CAT(0) cube complexes we would have that all of them are virtually special. It turns out that an even stronger statement holds.

Relying on work of Kahn and Markovic [KM12], Bergeron and Wise show in [BW12] that hyperbolic 3-manifold groups can be cubulated. That is, they construct a CAT(0) cube complex on which these groups act geometrically. This cubulation is obtained using Sageev's method [Sag95] via hyperplanes and half-space systems as introduced in Sect. 5.1. The half-space system arises from the many immersed, π_1-injective surfaces inside such a 3-manifold.

Combining Theorem 7.82 with what we just discussed, one obtains that every fundamental group of a closed 3-manifold M^3 is virtually special. Earlier work of Agol [Ago08] implies that if $\pi_1(M)$ is virtually special cubulated, then M virtually fibers. This means that M has a finite sheeted cover that admits a projection onto $\mathbb{S}^1$ whose pre-images of points are surfaces inside M, i.e. the fibers of this projection are surfaces.

So in total Agol's theorem provides an affirmative solution to Thurston's long-standing fibering conjecture, which claims that every 3-manifold group virtually fibers over the circle. Agol's theorem also provides a positive solution to the virtual Haken conjecture and implies that every such group has a finite sheeted cover that is a Haken manifold. Haken manifolds are particularly well understood as they can be decomposed into classifiable pieces.

Thurston's conjectures appeared as questions in his beautiful article [Ota14] titled "Three-dimensional manifolds, Kleinian groups and hyperbolic geometry". In this paper a list of 24 questions concerning the structure of 3-manifold groups were given, including famous problems like the geometrization conjecture. The positive solution to the virtual fibration and Haken conjecture conclude Thurston's program to some extend.

This does, however, not mean that there is noting else to explore in this direction. Agol [Ago12] and Wise [Wis14] both state several open problems. To name one: Given a compact, non-positively curved cube complex. Is there an algorithm that decides whether or not this complex is special?

The complete proof of Agol's theorem is too involved for the scope of these lecture notes. More details and further explanations of its strategy can be found in the following references:

There is a great blog entry at *Geometry and the imagination* by Danny Calegari from March 26th 2012 titled *Agol's virtual Haken Theorem* [Cal12]. Bestvina [Bes14] has written a piece in the Bulletin of the AMS. Calegari [Cal13] has Notes on Agol's virtual Haken theorem from a course given at UChicago in 2013. Another good source to read about Agol's work are Shepherds notes [She21b], which provide a more detailed account of Agol's original paper [Ago13]. Finally one should watch the recording of Agol's ICM talk [Ago12] on the topic.

More on special cube complexes can be found in Kropholler's expository article [Kro18] in Collection [KLMPN18] and also in Wilton's lecture notes [Wil17].

7.4 Phylogenetic Trees

In this section we discuss an application of cube complexes outside pure mathematics. The main goal is the following: Study and graphically illustrate hierarchical connections between different species given an uncertainty of the input data or in the process of creation of the data.

We introduce labeled N-trees which model the evolution of species or languages. A space of n-trees with varying side lengths is constructed and we will prove that this space carries the natural structure of a CAT(0) cube complex. The material in this section is based on the work of Billera, Holmes and Vogtmann [BHV01].

Fig. 7.7 Identical metric
labeled 4-trees

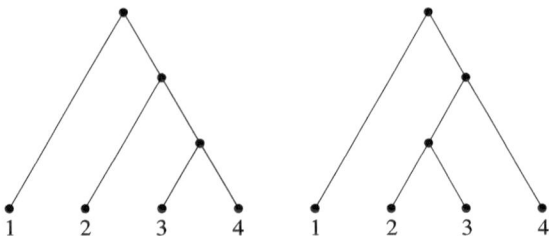

Fig. 7.7 Identical metric
labeled 4-trees

7.4.1   Modeling Mutation Using Trees

Suppose we have collected data on e.g. the genetic evolution of a species or the
transformation of a language over time. Often such data is obtained using statistical
methods. Suppose further that we now want to measure the mutation of the species
(or language) with the goal to develop models and graphical displays of their
evolution. We face the following problem: Different methods to evaluate the data
often lead to different outcomes. So the obtained models and graphical results differ
depending on the method used. This is not surprising, as it is often not entirely clear
how evolution went. It can, for example, be hard to measure how languages differ
and whether or not two of them have common ancestors.

Following [BHV01] we use simplicial trees equipped with some extra structure,
such as roots, leaves, etc,.., to represent the evolution of a set of species.

Definition 7.83 (Labeled n-Trees) A *rooted tree* is a simplicial tree T together
with a fixed vertex of valency ≥ 2, called its *root*. An *n-tree* is a rooted tree, which
has n vertices of valency 1, called *leafs*, and in which all vertices that are neither the
root nor a leaf have valency ≥ 3. A vertex of an n-tree is called *interior* if it is not a
leaf. An edge called *interior* if it does not contain a leaf. A vertex u is the *ancestor*
of a vertex v if u is adjacent to v and the unique shortest edge-path from the root to
v contains the edge between u and v. A *labeling* of an n-tree T is a bijection from
$\{1, 2, \ldots, n\}$ to the set of leafs of T. A tree T together with a fixed labeling is a
labeled tree.

In this section a metric realization of a (simplicial, n-) tree T, as given in
Definition 2.6, is called a *metric tree*. Such a metric tree is obtained from a simplicial
tree together with a map $l : E \to [0, \infty)$ that assigns to each of its edges a non-
negative number representing the length of the respective edge.

Example 7.84 We do not distinguish different ways of drawing a given simplicial
or metric tree. In Fig. 7.7, for example, both sides show the same metric labeled tree
illustrated in two different ways. The tree has four leafs, i.e. vertices adjacent to
only one edge, and is hence a 4-tree. The lengths of the edges in this metric tree are
drawn to scale.

Traditionally the development of species is modeled using a single mutation or
evolutionary species tree. This tree encodes the development of a species over time

and is also referred to as a phylogenetic tree. In modeling such a tree species, which share the most characteristics become classified as "siblings" and are assumed to share a common genetic ancestor. More formally, a (combinatorial) rooted tree is interpreted in the following way.

Notation 7.85 (Mutation Tree) Let T be a (combinatorial) rooted tree. Then

- each leaf represents a living species,
- all other nodes represent extinct (or sometimes also living) species, and
- nodes with a common ancestor (in the graph theoretic sense) correspond to species with common (genetic) ancestors.

If one wants to encode more information than just whether or not two species have a common ancestor one can consider metric trees with different edge lengths. Lengths of edges in a metric tree then encode additional information, such as time needed to branch, amount of genes different from the direct ancestor, number of mutations needed, etc.

In choosing more general labels, e.g. vectors in $\mathbb{R}^n$ or arbitrary sets, it would be possible to represent with one label several aspects in which the species differ. This more general approach is not considered in this chapter.

As already mentioned, one of the problems in constructing a phylogenetic tree is a certain degree of uncertainty about the precise connection between species and their development. So it is not always clear which n-tree provides the best model. One has to deal with multiple phylogenetic trees at a time, interpolate between them or estimate how likely it is for a tree to be a good model. To summarize, the main problems are:

Question 7.86 Consider the above tree model for the mutation of species as described in Notation 7.85. How can one

- interpolate between two trees or measure their distance?
- calculate the probability of appearance of a given set of trees in the space of all possible trees?

We later define a space of all labeled rooted trees in which these problems can be solved. First we explore how many interior edges an n-tree can have.

7.4.2 Exploring the Structure of n-Trees

The goal of this subsection is to collect some further information on the structure and number of *n*-trees. We start by exploring how to obtain $(n + 1)$-trees from *n*-trees.

Example 7.87 One can create new $(n + 1)$-trees from *n*-trees by attaching new edges. Have a look at Fig. 7.8. On the left there is a labeled rooted 6-tree. On the right hand side this tree has been enlarged (in two ways) to a labeled rooted 7-tree.

One can do this by either adding two new edges at a leaf. In the example two edges are added to leaf number 5, which turns leaf 5 into an interior vertex and produces two new leafs (labeled 5 and 7). Or one can add one edge to an interior vertex and produce a new leaf (numbered 7) as shown in the picture in the lower right-hand corner. Here an edge has been added to the ancestor of 1,2,3 and 4.

Another question is how to produce new n-trees from a given n-tree, that is changing the tree structure while keeping the number of ends intact.

Example 7.88 In Fig. 7.9 an example is shown for how to produce a new 5-tree from a given one. By adding a new interior edge at a vertex of valence four (or higher) the number of leafs stays the same while the number of interior edges increases by one. This can be done (with the same combinatorial tree pictured) in three ways. One may either number the leafs from left to right by 1,2,3,4,5 as shown in the tree on the right or, alternatively by 2,1,3,4,5 or 3,1,2,4,5. In total there are three different labeled rooted trees on the same underlying combinatorial tree.

The manipulations discussed in the example play a crucial role in the proof of the next proposition. It is not hard to see that they are the only ways in which such a tree can be altered. Recall that a *binary tree* is a rooted tree where all interior vertices (except for the root) are contained in three edges.

Proposition 7.89 *An n-tree has at most $n - 2$ interior edges where equality holds for binary trees.*

Proof The proof is by induction on the number n of leafs. For $n = 2$ there is only one possible tree with no interior edges. Every n-tree arises from an $(n - 1)$-tree by adding an additional leaf. As we have illustrated in Fig. 7.8 a leaf can be attached in two ways. Either one glues two new edges to an existing leaf or one glues a

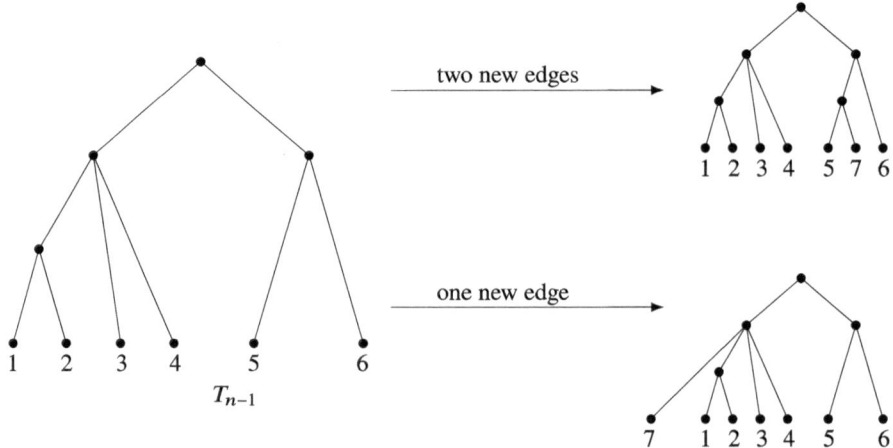

Fig. 7.8 Constructing an n-tree from an $(n - 1)$-tree

Fig. 7.9 Enlarge the labeled
5-tree on the left to the binary
tree on the right by adding a
new edge at the encircled
vertex

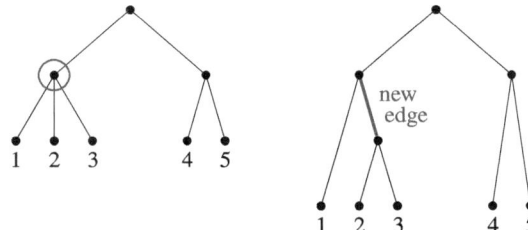

single new edge to a vertex of degree strictly bigger than one. In the first case one
obtains one additional interior edge. No new interior edge appears in the second
case where one new edge is glued to a non-leaf. By induction we hence have at
most $(n-1) + 1 - 2 = n - 2$ interior edges in every n-tree.

To prove that binary trees attain the maximal number of interior vertices we need
to consider an additional method of enlarging the number of interior edges of a
rooted tree without increasing the number of leafs.

Proceed as follows: find a vertex v of degree at least four. One edge at such a
vertex is closer to the root as the others. We say it *faces up* while we refer to the
others as edges *facing down*. Take two edges e_1, e_2 that are connected to v and
facing down. Cut the tree apart at v and replace v by three vertices. Refer to the
vertex still attached to the edge facing up again by v and let v_i be the two new
copies of v attached to the two edges e_i facing down. Now glue in a new edge e
between v on one end and the two endpoints v_i of $e_i, i = 1, 2$ on the other end.
Doing so, the number of leafs stays the same while the number of interior edges
was increased by one. This process is illustrated in Fig. 7.9 where v is the encircled
vertex in the tree on the left.

This method is applicable if and only if there exists a vertex of degree strictly
bigger than three as otherwise this would produce an interior vertex of degree two.
This is not allowed in our trees. Hence, binary trees can not be enlarged without
adding leafs as all their interior vertices have degree three. And thus binary trees
have the maximal number of interior edges. □

From Proposition 7.89 we obtain that the binary trees are the extreme cases when
it comes to number of interior edges. One can also count how many different ones
exist. This question was already considered by Schröder in the nineteenth century
[Sch70].

Proposition 7.90 *There exist* $(2n - 3)!! = \frac{(2n-2)!}{2^{n-2}(n-1)!}$ *binary labeled n-trees.*

7.4.3 The Space of Phylogenetic Trees

As already mentioned in the first subsection we now turn the collection of all n-trees
into a space $\mathcal{T}_n$. The construction is done as follows.

Fig. 7.10 The Petersen
graph. This is the shape of the
link of the origin in $\mathcal{T}_4$

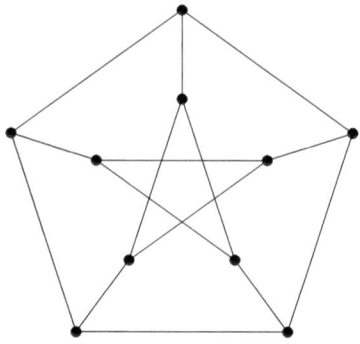

Definition 7.91 (The Space of Phylogenetic Trees) Let T be a binary labeled n-tree with interior edges enumerated by $e_1, e_2, \ldots, e_{n-2}$. Associate to T a quadrant $Q_T \cong [0, \infty)^{n-2}$ where each point $p = (p_1, p_2, \ldots, p_{n-2})^t$ in Q_T corresponds to a metric tree having underlying combinatorial tree T. An interior edge e_i has length p_i for all i. The faces of Q_T correspond to non-binary n-trees where certain interior edges are collapsed to have length 0.

Define the *space of phylogenetic n-trees* $\mathcal{T}_n$ to be the quotient

$$\mathcal{T}_n := \left(\bigcup_{T \in \mathcal{B}_n} Q_T \right) / \sim.$$

Here $\mathcal{B}_n$ stands for the set of binary n-trees and we denote by $\sim$ the equivalence relation on points in quadrants induced by equivalence of labeled metric n-trees. That is, points $p \in Q_T$ and $p' \in Q_{T'}$ are identified in the quotient if and only if p and p' represent the same labeled metric n-tree.

Let us look at small examples of the space of phylogenetic trees.

Example 7.92 The set $\mathcal{B}_3$ contains $(2 \cdot 3 - 3)!! = 3$ labeled binary 3-trees. Hence, $\mathcal{T}_3$ contains three quadrants of dimension one (which are rays) glued together at the origin.

The set $\mathcal{B}_4$ contains fifteen elements and each triple defines a 2-dimensional quadrant. At each face of dimension one three such quadrants meet. One can show, compare Exercise 7.105, that the link of the origin in $\mathcal{T}_4$ is the Petersen graph. A picture of this graph is shown in Fig. 7.10

Note that there are many embeddings of $\mathcal{T}_3$ into $\mathcal{T}_4$ and, in general, of $\mathcal{T}_k$ into $\mathcal{T}_n$ for arbitrary integers $k < n$. We end this section by turning $\mathcal{T}_n$ into a CAT(0) cube complex.

Theorem 7.93 (The Phylogenetic Trees Form a Cube Complex) *For all $n \in \mathbb{N}$ the space $\mathcal{T}_n$ carries the structure of a CAT(0) cube complex.*

Proof As $\mathcal{T}_n$ is built from Euclidean quadrants there is an obvious way to define a metric d on $\mathcal{T}_n$. Let d be the unique length metric such that on each quadrant the restriction $d|_{Q_T}$ equals the restricted Euclidean metric. The cube complex structure on every quadrant is obtained by taking unit cubes on integral vertices in $\mathbb{R}^n$. The gluing of quadrants then respects the cubical structure by construction.

It is not hard to check that the Petersen graph is a CAT(1) space. We may thus apply Theorem 3.33 to the link $Y = \mathrm{lk}_{\mathcal{T}_n}(0)$ and obtain that the cone over Y is CAT(0). □

This theorem has some nice consequences: In a CAT(0) cube complex distances between vertices are easy to measure. Geodesics in a CAT(0) cube complex are algorithmically well understood and can effectively be computed. Additional structure, such as measures, may be put on this space, which allows to interpolate between trees and to compute likelihood of appearance.

7.5 Exercises

Exercise 7.94 Discuss: Why is the assumption that the action of G is simplicial relevant in the setting of Sect. 7.1?

Exercise 7.95 Let G be a finitely generated group with generating set $\{g_1, \ldots, g_k\}$ and suppose that G acts isometrically on X. Show that if $\mathrm{Fix}_\Phi(\langle g_i \rangle) \neq \emptyset$ and $\mathrm{Fix}_\Phi(\langle g_i \rangle) \cap \mathrm{Fix}_\Phi(\langle g_j \rangle) \neq \emptyset$ for all $i, j \in \{1, \ldots, k\}$, then $\mathrm{Fix}_\Phi(G) \neq \emptyset$. That is, Corollary 7.8 is true without the assumption that the action is without inversions.

Exercise 7.96 Let X be a CAT(0) cube complex. Show that the metric d_1 on the vertices of X introduced in the proof of Theorem 5.15 and the metric d from Definition 4.26 agree.

Exercise 7.97 Prove the statements in Remark 7.38.

Exercise 7.98 Show that every subgroup of a RAAG is the fundamental group of a special cube complex. Is such a special cube complex always compact?

Exercise 7.99 Draw a special cube complex with at least four hyperplanes and its hyperplane complex. Does there exist a special cube complex with a complete graph on n vertices as its hyperplane graph? Try for small $n \leq 5$ first.

Exercise 7.100 Prove that the product $X_1 \times X_2$ of a pair of special cube complexes X_1 and X_2 is again special. (See also Corollary 3.6 in [HW08].)

Exercise 7.101 Show that the universal cover of every special cube complex is special.

If you're up for a challenge then the next exercise is for you. A proof of this fact can be found in [She21b] or [Sag95].

Exercise 7.102 Prove that in fact every CAT(0) cube complex is special.

Fig. 7.11 Surface of genus
two

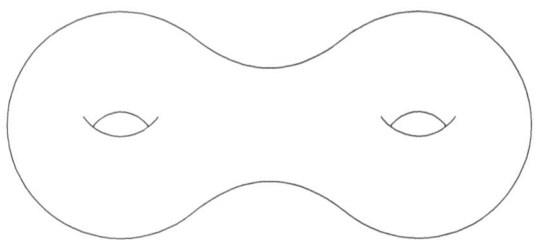

Exercise 7.103 Prove Lemma 7.70. More precisely: Let X be a cube complex and use notation as in Notation 4.47. Show that X is locally special if and only if the following two conditions are satisfied for all $3 \leq k \in \mathbb{N}$:

1. For any edge e containing a vertex v_k any k-corner containing e has a k-corner adjacent along e. This k-corner is unique and equals $\sigma_e(c)$.
2. For any two edges e, e' containing v_k and any k-corner $c : C_k \to X$ we have
   $\sigma_e(\sigma_{e'}(c)) = \sigma_{e'}(\sigma_e(c))$.

Exercise 7.104 Let S be a surface of genus two as shown in Fig. 7.11.

1. Divide S symmetrically into eight right-angled hyperbolic five-gons and describe the dual polyhedral complex $S^\vee$ of this partition. What are the cells of $S^\vee$?
2. Describe the vertex links of the polyhedral complex $S^\vee$ and decide whether $S^\vee$ is locally CAT(0).
3. Describe the hyperplanes of $S^\vee$ using the partition of S.
4. Convince yourself that the following are true for all hyperplanes H and K in $S^\vee$:

   (a) H is not self-intersecting.
   (b) H is not directly self-osculating.
   (c) H and K do not inter-osculate.

5. Which of the previous statements are still true for coverings of S?

Exercise 7.105 Prove that the link of the origin in $\mathcal{T}_4$ is the Petersen graph.

Exercise 7.106 Figure out and prove in how many ways $\mathcal{T}_3$ can be embedded into $\mathcal{T}_4$? Can you guess how many ways there are to embed $\mathcal{T}_k$ into $\mathcal{T}_n$ for arbitrary integers $k < n$?

References

[AB08]   P. Abramenko, K.S. Brown, *Buildings, Theory and Applications. Graduate Texts in Mathematics*, vol. 248 (Springer, New York, 2008)

[ABY14]   F. Ardila, T. Baker, R. Yatchak, Moving robots efficiently using the combinatorics of CAT(0) cubical complexes. SIAM J. Discrete Math. **28**(2), 986–1007 (2014)

[AG04]   A. Abram, R. Ghrist, State complexes for metamorphic robots. Int. J. Robot. Res. **23**(7–8), 811–826 (2004)

[AGM09]   I. Agol, D. Groves, J.F. Manning, Residual finiteness, QCERF and fillings of hyperbolic groups. Geom. Topol. **13**(2), 1043–1073 (2009)

[Ago08]   I. Agol, Criteria for virtual fibering. J. Topol. **1**(2), 269–284 (2008)

[Ago12]   I. Agol, Virtual properties of 3-manifolds (recording of icm talk) (2012). https://www.youtube.com/watch?v=Y6ACJYvfs5U

[Ago13]   I. Agol, The virtual Haken conjecture (with an appendix by Ian Agol, Daniel Groves and Jason Manning). Doc. Math. **18**, 1045–1087 (2013)

[All21]   D. Allcock, Most big mapping class groups fail the Tits alternative. Algebr. Geom. Topol. **21**(7), 3675–3688 (2021)

[AM20]   F. Ardila-Mantilla, CAT(0) geometry, robots, and society. Not. Am. Math. Soc. **67**(7), 977–987 (2020)

[AOS12]   F. Ardila, M. Owen, S. Sullivant, Geodesics in CAT(0) cubical complexes. Adv. Appl. Math. **48**(1), 142–163 (2012)

[Bal95]   W. Ballmann, *Lectures on Spaces of Nonpositive Curvature*. DMV Seminar, vol. 25 (Birkhäuser Verlag, Basel, 1995). With an appendix by Misha Brin

[BB05]   A. Björner, F. Brenti, *Combinatorics of Coxeter Groups*, vol. 231 (Springer, New York, 2005)

[BB10]   A.V. Borovik, A. Borovik, *Mirrors and Reflections. The Geometry of Finite Reflection Groups*. Universitext (Springer, New York, 2010)

[BBI01]   D. Burago, Y. Burago, S. Ivanov, *A Course in Metric Geometry* Graduate Studies in Mathematics, vol. 33 (American Mathematical Society, Providence, 2001)

[BC08]   H.-J. Bandelt, V. Chepoi, Metric graph theory and geometry: a survey, in *Surveys on Discrete and Computational Geometry. Twenty Years Later. AMS-IMS-SIAM Summer Research Conference*, Snowbird, UT, June 18–22, 2006 (American Mathematical Society, Providence, 2008), pp. 49–86

[Ber83]   V.N. Berestovskiĭ, Borsuk's problem on the metrization of a polyhedron. Sov. Math. Dokl. **27**, 56–59 (1983)

[Bes14]   M. Bestvina, Geometric group theory and 3-manifolds hand in hand: the fulfillment of Thurston's vision. Bull. Am. Math. Soc. New Ser. **51**(1), 53–70 (2014)

© The Author(s), under exclusive license to Springer Nature Switzerland AG 2023   173
P. Schwer, *CAT(0) Cube Complexes*, Lecture Notes in Mathematics 2324,
https://doi.org/10.1007/978-3-031-43622-2

[BF22]   J. Beyrer, E. Fioravanti, Cross ratios and cubulations of hyperbolic groups. Math. Ann. **384**(3–4), 1547–1592 (2022)
[BFH00]   M. Bestvina, M. Feighn, M. Handel, The Tits alternative for Out(F_n). I: dynamics of exponentially-growing automorphisms. Ann. Math. (2) **151**(2), 517–623 (2000)
[BFIM21]   J. Beyrer, E. Fioravanti, M. Incerti-Medici, CAT(0) cube complexes are determined by their boundary cross ratio. Groups Geom. Dyn. **15**(1), 313–333 (2021)
[BGS85]   W. Ballmann, M. Gromov, V. Schroeder, *Manifolds of Nonpositive Curvature*. Progress in Mathematics, vol. 61 (Birkhäuser Boston Inc., Boston, 1985)
[BH99]   M.R. Bridson, A. Haefliger, *Metric Spaces of Non-positive Curvature*, Grundlehren der Mathematischen Wissenschaften, vol. 319 (Springer, Berlin, 1999)
[BHS17]   J. Behrstock, M.F. Hagen, A. Sisto, Hierarchically hyperbolic spaces. I: curve complexes for cubical groups. Geom. Topol. **21**(3), 1731–1804 (2017)
[BHV01]   L.J. Billera, S.P. Holmes, K. Vogtmann, Geometry of the space of phylogenetic trees. Adv. Appl. Math. **27**(4), 733–767 (2001)
[BHW11]   N. Bergeron, F. Haglund, D.T. Wise, Hyperplane sections in arithmetic hyperbolic manifolds. J. Lond. Math. Soc. II. Ser. **83**(2), 431–448 (2011)
[BM00]   T. Brady, J.P. McCammond, Three-generator Artin groups of large type are biautomatic. J. Pure Appl. Algebra **151**(1), 1–9 (2000)
[BN14]   V.N. Berestovskiĭ, Y.G. Nikonorov, Generalized normal homogeneous Riemannian metrics on spheres and projective spaces. Ann. Global Anal. Geom. **45**(3), 167–196 (2014)
[Bou02]   N. Bourbaki, *Lie Groups and Lie Algebras*, Chapters 4–6. Elements of Mathematics (Berlin) (Springer, Berlin, 2002). Translated from the 1968 French original by Andrew Pressley
[Bri07]   M.R. Bridson, Helly's theorem, CAT(0) spaces, and actions of automorphism groups of free groups, preprint (2007)
[Bri08]   M.R. Bridson, A condition that prevents groups from acting nontrivially on trees, in *The Zieschang Gedenkschrift*. Geom. Topol. Monogr., vol. 14 (Geom. Topol. Publ., Coventry, 2008), pp. 129–133
[Bri12]   M.R. Bridson, On the dimension of CAT(0) spaces where mapping class groups act. J. Reine Angew. Math. **673**, 55–68 (2012)
[Bro89]   K.S. Brown, *Buildings* (Springer, New York, 1989)
[BS85]   M.G. Brin, C.C. Squier, Groups of piecewise linear homeomorphisms of the real line. Invent. Math. **79**, 485–498 (1985)
[BŚ99]   W. Ballmann, J.Świątkowski, On groups acting on nonpositively curved cubical complexes. Enseign. Math. (2) **45**(1–2), 51–81 (1999)
[BT84]   F. Bruhat, J. Tits, Groupes réductifs sur un corps local. II. Schémas en groupes. Existence d'une donnée radicielle valuée. Inst. Hautes Études Sci. Publ. Math. **60**, 197–376 (1984)
[BW12]   N. Bergeron, D.T. Wise, A boundary criterion for cubulation. Am. J. Math. **134**(3), 843–859 (2012)
[BZKL22]   M. Ben-Zvi, R. Kropholler, R.A. Lyman, Folding-like techniques for CAT(0) cube complexes. Math. Proc. Camb. Philos. Soc. **173**(1), 227–238 (2022)
[Cal12]   D. Calegari, Geometry and the imagination: Agol's virtual Haken theorem. Last accessed 17 Feb 2023 (2012). https://lamington.wordpress.com/tag/virtual-haken-conjecture/
[Cal13]   D. Calegari, Cube complexes (2013). https://math.uchicago.edu/~dannyc/courses/agol_virtual_haken/agol_notes.pdf
[CD95]   R. Charney, M.W. Davis, Finite $K(\pi, 1)$s for Artin groups, in *Prospects in Topology. Proceedings of a Conference in Honor of William Browder*, Princeton, March 1994 (Princeton University Press, Princeton, 1995), pp. 110–124
[CDH10]   I. Chatterji, C. Druţu, F. Haglund, Kazhdan and Haagerup properties from the median viewpoint. Adv. Math. **225**(2), 882–921 (2010)

[CFP96]   J.W. Cannon, W.J. Floyd, W.R. Parry, Introductory notes on Richard Thompson's groups. Enseign. Math. (2) **42**(3–4), 215–256 (1996)

[Che00]   V. Chepoi, Graphs of some CAT(0) complexes. Adv. Appl. Math. **24**(2), 125–179 (2000)

[CM05]   P.-E. Caprace, B. Mühlherr, Reflection triangles in Coxeter groups and biautomaticity. J. Group Theory **8**(4), 467–489 (2005)

[CN05]   I. Chatterji, G. Niblo, From wall spaces to CAT(0) cube complexes. Int. J. Algebra Comput. **15**(5–6), 875–885 (2005)

[CS11]   P.-E. Caprace, M. Sageev, Rank rigidity for CAT(0) cube complexes. Geom. Funct. Anal. **21**(4), 851–891 (2011)

[Dav98]   M.W. Davis, Buildings are CAT(0), in *Geometry and Cohomology in Group Theory (Durham, 1994)*. London Math. Soc. Lecture Note Ser., vol. 252 (Cambridge University Press, Cambridge, 1998), pp. 108–123

[DJ00]   M.W. Davis, T. Januszkiewicz, Right-angled Artin groups are commensurable with right-angled Coxeter groups. J. Pure Appl. Algebra **153**(3), 229–235 (2000)

[DL21]   P. Dani, I. Levcovitz, Subgroups of right-angled Coxeter groups via stallings-like techniques. J. Comb. Algebra **5**(3), 237–295 (2021)

[dlH00]   P. de la Harpe, *Topics in Geometric Group Theory* (The University of Chicago Press, Chicago, 2000)

[DM99]   M.W. Davis, G. Moussong, Notes on nonpositively curved polyhedra, in *Low Dimensional Topology (Eger, 1996/Budapest, 1998)*. Bolyai Soc. Math. Stud., vol. 8 (János Bolyai Math. Soc., Budapest, 1999), pp. 11–94

[Dro87]   C. Droms, Isomorphisms of graph groups. Proc. Am. Math. Soc. **100**, 407–408 (1987)

[DS00]   M.J. Dunwoody, E.L. Swenson, The algebraic torus theorem. Invent. Math. **140**(3), 605–637 (2000)

[Far09]   B. Farb, Group actions and Helly's theorem. Adv. Math. **222**(5), 1574–1588 (2009)

[Fio18]   E. Fioravanti, The Tits alternative for finite rank median spaces. Enseign. Math. **64**(1–2), 89–126 (2018)

[FLS22]   E. Fioravanti, I. Levcovitz, M. Sageev, Coarse cubical rigidity, preprint, arXiv:2210.11418 (2022)

[Fre45]   H. Freudenthal, Über die Enden diskreter Räume und Gruppen. Comment. Math. Helv. **17**, 1–38 (1945)

[GdlH90]   E. Ghys, P. de la Harpe (eds.), *Sur les groupes hyperboliques d'après Mikhael Gromov (On the Hyperbolic Groups à la M. Gromov)*. Progress in Mathematics, vol. 83 (Birkhäuser, Boston, 1990), vii, 285 p. sFr. 68.00

[Ger98]   V. Gerasimov, Fixed-point-free actions on cubings [translation of algebra, geometry, analysis and mathematical physics (Russian) (Novosibirsk, 1996), 91–109, 190, Izdat. Ross. Akad. Nauk Sibirsk. Otdel. Inst. Mat., Novosibirsk, 1997; MR1624115 (99c:20049)]. Siberian Adv. Math. **8**(3), 36–58 (1998)

[Gro87]   M. Gromov, Hyperbolic groups, in *Essays in Group Theory* (Springer, Berlin, 1987), pp. 75–263

[Hae21]   T. Haettel, Virtually cocompactly cubulated Artin-Tits groups. Int. Math. Res. Not. IMRN **4**, 2919–2961 (2021)

[Hae22]   T. Haettel, Cubulation of some triangle-free Artin groups. Groups Geom. Dyn. **16**(1), 287–304 (2022)

[Hag13]   M.F. Hagen, The simplicial boundary of a CAT(0) cube complex. Algebr. Geom. Topol. **13**(3), 1299–1367 (2013)

[Hag18]   M.F. Hagen, Corrigendum to: "The simplicial boundary of a CAT(0) cube complex". Algebr. Geom. Topol. **18**(2), 1251–1256 (2018)

[HJP16]   J. Huang, K. Jankiewicz, P. Przytycki, Cocompactly cubulated 2-dimensional Artin groups. Comment. Math. Helv. **91**(3), 519–542 (2016)

[HK18]   J. Huang, B. Kleiner, Groups quasi-isometric to right-angled Artin groups. Duke Math. J. **167**(3), 537–602 (2018)

[HNN50]   G. Higman, B.H. Neumann, H. Neumann, Embedding theorems for groups. J. Lond. Math. Soc. **24**, 247–254 (1950)

[Hop44]   H. Hopf, Enden offener Räume und unendliche diskontinuierliche Gruppen. Comment. Math. Helv. **16**, 81–100 (1944)

[Hua18]   J. Huang, Commensurability of groups quasi-isometric to RAAGs. Invent. Math. **213**(3), 1179–1247 (2018)

[Hum72]   J.E. Humphreys, *Introduction to Lie Algebras and Representation Theory*. Graduate Texts in Mathematics, vol. 9 (Springer, New York, 1972)

[HW08]   F. Haglund, D.T. Wise, Special cube complexes. Geom. Funct. Anal. **17**(5), 1551–1620 (2008)

[HW15]   M.F. Hagen, D.T. Wise, Cubulating hyperbolic free-by-cyclic groups: the general case. Geom. Funct. Anal. **25**(1), 134–179 (2015)

[HW16]   M.F. Hagen, D.T. Wise, Cubulating hyperbolic free-by-cyclic groups: the irreducible case. Duke Math. J. **165**(9), 1753–1813 (2016)

[Iva84]   N.V. Ivanov, Algebraic properties of the Teichmüller modular group. Sov. Math. Dokl. **29**, 288–291 (1984)

[Kle99]   B. Kleiner, The local structure of length spaces with curvature bounded above. Math. Z. **231**(3), 409–456 (1999)

[KLMPN18]   P.H. Kropholler, I.J. Leary, C. Martínez-Pérez, B.E.A. Nucinkis (eds.) *Geometric and Cohomological Group Theory. Proceedings of the London Mathematical Society Durham Symposium*, Durham, August 12–22, 2013. *Lond. Math. Soc. Lect. Note Ser.*, vol. 444 (Cambridge University Press, Cambridge, 2018)

[KM12]   J. Kahn, V. Markovic, Immersing almost geodesic surfaces in a closed hyperbolic three manifold. Ann. Math. (2) **175**(3), 1127–1190 (2012)

[Kro18]   R.P. Kropholler, Special cube complexes (based on lectures of Piotr Przytycki), in *Geometric and Cohomological Group Theory. Proceedings of the London Mathematical Society Durham Symposium*, Durham, August 12–22, 2013, pp. 46–66 (Cambridge University Press, Cambridge, 2018)

[Lö17]   C. Löh, *Geometric Group Theory. An Introduction*. Universitext (Springer, Berlin, 2017)

[Lea13]   I.J. Leary, A metric Kan-Thurston theorem. J. Topol. **6**(1), 251–284 (2013)

[Led20]   N.J. Leder, *Automorphism Groups of Graph Products and Serre's Property FA*. Univ. Münster, Mathematisch-Naturwissenschaftliche Fakultät, Münster. Fachbereich Mathematik und Informatik (Diss.) (2020)

[Lev21]   I. Levcovitz, Comparing the Roller and $B(X)$ boundaries of CAT(0) cube complexes. Israel J. Math. **242**(1), 129–170 (2021)

[LS01]   R.C. Lyndon, P.E. Schupp, *Combinatorial Group Theory* (Springer, Berlin, 2001)

[McC85]   J. McCarthy, A Tits-alternative for subgroups of surface mapping class groups. Trans. Am. Math. Soc. **291**, 583–612 (1985)

[Mei08]   J. Meier, *Groups, Graphs and Trees*. London Mathematical Society Student Texts, vol. 73 (Cambridge University Press, Cambridge, 2008)

[Mil68]   J.W. Milnor, A note on curvature and fundamental group. J. Differ. Geom. **2**, 1–7 (1968)

[Min12]   A. Minasyan, Hereditary conjugacy separability of right-angled Artin groups and its applications. Groups Geom. Dyn. **6**(2), 335–388 (2012)

[MP20]   A. Martin, P. Przytycki, Tits alternative for Artin groups of type FC. J. Group Theory **23**(4), 563–573 (2020)

[Mug21]   D. Mugnolo, What is actually a metric graph? (2021). arXiv:1912.07549, math.CO

[Nib04]   G.A. Niblo, A geometric proof of Stallings' theorem on groups with more than one end. Geom. Dedicata **105**, 61–76 (2004)

[Nic04]   B. Nica, Cubulating spaces with walls. Algebr. Geom. Topol. **4**, 297–309 (2004)

[NR98]   G.A. Niblo, M.A. Roller, Groups acting on cubes and Kazhdan's property (T). Proc. Am. Math. Soc. **126**(3), 693–699 (1998)

[NR03]   G.A. Niblo, L.D. Reeves, Coxeter groups act on CAT(0) cube complexes. J. Group Theory **6**(3), 399–413 (2003)

[NS13]   A. Nevo, M. Sageev, The Poisson boundary of CAT(0) cube complex groups. Groups Geom. Dyn. **7**(3), 653–695 (2013)

[OPM21]   D. Osajda, P. Przytycki, J. McCammond, Tits alternative for groups acting properly on 2-dimensional recurrent complexes. Adv. Math. **391**, 22 (2021)

[Ota14]   J.-P. Otal, William P. Thurston: "Three-dimensional manifolds, Kleinian groups and hyperbolic geometry". Jahresber. Dtsch. Math.-Ver. **116**(1), 3–20 (2014)

[OW11]   Y. Ollivier, D.T. Wise, Cubulating random groups at density less than 1/6. Trans. Am. Math. Soc. **363**(9), 4701–4733 (2011)

[PW14]   P. Przytycki, D.T. Wise, Graph manifolds with boundary are virtually special. J. Topol. **7**(2), 419–435 (2014)

[Rol98]   M. Roller, Poc sets, median algebras and group actions (1998). Habilitationsschrift, arXiv:1607.07747, [math.GN]

[Rol12]   P. Rolli, Notes on CAT(0) cube complexes, preprint (2012)

[Ron89]   M.A. Ronan, *Lectures on Buildings*. Perspectives in Mathematics, vol. 7 (Academic Press Inc., Boston, 1989)

[RS22]   Y.S. Rego, P. Schwer, The galaxy of Coxeter groups to appear in the Journal of Algebra, https://arxiv.org/abs/2211.17038 (2022)

[Sag95]   M. Sageev, Ends of group pairs and non-positively curved cube complexes. Proc. Lond. Math. Soc. (3) **71**(3), 585–617 (1995)

[Sag14]   M. Sageev, CAT(0) cube complexes and groups, in *Geometric Group Theory*. Lecture Notes from the IAS/Park City Mathematics Institute (PCMI) Graduate Summer School, Princeton, 2012 (American Mathematical Society, Providence; Institute for Advanced Study (IAS), Princeton, 2014), pp. 7–54

[Sch70]   E. Schröder, 4 combinatorische Probleme. Schlömilch Z. **15**, 361–376 (1870)

[Sch19]   P. Schwer, Lecture notes on CAT(0) cubical complexes (2019). AMS Open Math Notes, OMN:201907.110800, OMN:201907.110800

[Sch22]   P. Schwer, Shadows in the wild-folded galleries and their applications. Jahresber. Dtsch. Math.-Ver. **124**(1), 3–41 (2022)

[Ser03]   J.-P. Serre, *Trees. Transl. from the French by John Stillwell* (Springer, Berlin, 2003)

[She21a]   S. Shepherd, Agol's theorem on hyperbolic cubulations. Rocky Mt. J. Math. **51**(3), 1037–1073 (2021)

[She21b]   S. Shepherd, Finite covers of graphs and cube complexes (2021). Available at https://ethos.bl.uk/OrderDetails.do?uin=uk.bl.ethos.833378

[Shv55]   A.S. Shvarts, A volume invariant of coverings. Dokl. Akad. Nauk SSSR **105**, 32–34 (1955)

[Sta68]   J.R. Stallings, On torsion-free groups with infinitely many ends. Ann. Math. (2) **88**, 312–334 (1968)

[SW05]   M. Sageev, D.T. Wise, The Tits alternative for CAT(0) cubical complexes. Bull. Lond. Math. Soc. **37**(5), 706–710 (2005)

[Tit72]   J. Tits, Free subgroups in linear groups. J. Algebra **20**, 250–270 (1972)

[Var14]   O. Varghese, Fixed points for actions of $\mathrm{Aut}(F_n)$ on CAT(0) spaces. Münster J. Math. **7**(2), 439–462 (2014)

[Var19]   O. Varghese, A condition that prevents groups from acting fixed point free on cube complexes. Geom. Dedicata **200**, 85–91 (2019)

[Vog02]   K. Vogtmann, Automorphisms of free groups and outer space. Geom. Dedicata **94**, 1–31 (2002)

[VSvRK21] A. Voigt, P. Schwer, N. von Rotberg, N. Knopf, Tricco v1.0.0 – a cubulation-based method for computing connected components on triangular grids. Geosci. Model Dev. Discuss. **2021**, 1–23 (2021)

[Wil17]   H. Wilton, Non-positively curved cube complexes. AMS Open Math Notes. OMN:201704.110697, pp. 1–41 (2017)

[Wis02]   D.T. Wise, The residual finiteness of negatively curved polygons of finite groups. Invent. Math. **149**(3), 579–617 (2002)

[Wis04]   D.T. Wise, Cubulating small cancellation groups. Geom. Funct. Anal. **14**(1), 150–214 (2004)

[Wis12]   D.T. Wise, *From Riches to Raags: 3-Manifolds, Right-Angled Artin Groups, and Cubical Geometry*, vol. 117 (American Mathematical Society, Providence; Conference Board of the Mathematical Sciences, Washington, 2012)

[Wis14]   D.T. Wise, The cubical route to understanding groups, in *Proceedings of the International Congress of Mathematicians (ICM 2014)*, Seoul, August 13–21, 2014. Vol. II: Invited lectures, pp. 1075–1099 (KM Kyung Moon Sa, Seoul, 2014)

[Wis21]   D.T. Wise, *The Structure of Groups with a Quasiconvex Hierarchy*. Annals of Mathematics Studies, vol. 209 (Princeton University Press, Princeton, 2021)

[Wri12]   N. Wright, Finite asymptotic dimension for CAT(0) cube complexes. Geom. Topol. **16**(1), 527–554 (2012)

[Xie04]   X. Xie, Tits alternative for closed real analytic 4-manifolds of nonpositive curvature. Topology Appl. **136**(1–3), 87–121 (2004)

Index

G-equivariant, 125
M_κ^n, 47

$A(\Gamma)$, 72
action, 17, 135
  Cayley graph, 22
  cocompact, 113, 126, 129
  on a cube complex, 151, 154
  cubical, 134
  equivariant, 125
  isometric, 18, 137
  limit point, 152
  polyhedral, 134
  properly discontinuous, 23, 125, 153, 154
  semisimple, 153
  simplicial, 134, 136, 171
  virtually solvable, 153
  without inversion, 134
adjacent corner, 160
Alexandrov's lemma, 24, 25, 38, 41, 43
algebraic torus theorem, 151
all-right spherical
  complex, 62–64
  face, 62
  shape, 62
amalgam (*see* amalgamated free product)
amalgamated free product, 144, 145, 151
  normal form, 147
  reduced expression, 147
  reduced form, 147
ancestor, 166
angle, 24, 25, 36
  Alexandrov angle, 38
  between geodesics, 37

comparison angle, 36
angle metric, 37
approximate midpoints, 44
arrow diagram, 105
Artin group
  right-angled, 157
A-special, 157
automorphism group, 136

ball
  closed, 12, 41
  open, 12, 53
barycentric subdivision, 58, 67, 74, 157
Bass–Serre tree, 145, 150, 155
Berestovskiĭ's theorem, 41, 44, 55
binary tree, 168, 170
boundary map, 86
Bridson's lemma, 153
Bruhat–Tits
  fixed point theorem, 34, 136
  theorem, 33

Cartan–Hadamard theorem, 35
CAT(0)
  angle metric, 37
  4-point condition, 44
  property, 32, 38
  space, 35, 38, 41, 44, 142, 170
    Alexandrov characterization, 38
    angle (*see* angle)
    convex set, 45
    Helly's theorem, 142
    link condition, 55

© The Author(s), under exclusive license to Springer Nature Switzerland AG 2023
P. Schwer, *CAT(0) Cube Complexes*, Lecture Notes in Mathematics 2324,
https://doi.org/10.1007/978-3-031-43622-2

LECTURE NOTES IN MATHEMATICS  Springer

Editors in Chief: J.-M. Morel, B. Teissier;

Editorial Policy

1. Lecture Notes aim to report new developments in all areas of mathematics and their applications – quickly, informally and at a high level. Mathematical texts analysing new developments in modelling and numerical simulation are welcome.

   Manuscripts should be reasonably self-contained and rounded off. Thus they may, and often will, present not only results of the author but also related work by other people. They may be based on specialised lecture courses. Furthermore, the manuscripts should provide sufficient motivation, examples and applications. This clearly distinguishes Lecture Notes from journal articles or technical reports which normally are very concise. Articles intended for a journal but too long to be accepted by most journals, usually do not have this "lecture notes" character. For similar reasons it is unusual for doctoral theses to be accepted for the Lecture Notes series, though habilitation theses may be appropriate.

2. Besides monographs, multi-author manuscripts resulting from SUMMER SCHOOLS or similar INTENSIVE COURSES are welcome, provided their objective was held to present an active mathematical topic to an audience at the beginning or intermediate graduate level (a list of participants should be provided).

   The resulting manuscript should not be just a collection of course notes, but should require advance planning and coordination among the main lecturers. The subject matter should dictate the structure of the book. This structure should be motivated and explained in a scientific introduction, and the notation, references, index and formulation of results should be, if possible, unified by the editors. Each contribution should have an abstract and an introduction referring to the other contributions. In other words, more preparatory work must go into a multi-authored volume than simply assembling a disparate collection of papers, communicated at the event.

3. Manuscripts should be submitted either online at www.editorialmanager.com/lnm to Springer's mathematics editorial in Heidelberg, or electronically to one of the series editors. Authors should be aware that incomplete or insufficiently close-to-final manuscripts almost always result in longer refereeing times and nevertheless unclear referees' recommendations, making further refereeing of a final draft necessary. The strict minimum amount of material that will be considered should include a detailed outline describing the planned contents of each chapter, a bibliography and several sample chapters. Parallel submission of a manuscript to another publisher while under consideration for LNM is not acceptable and can lead to rejection.

4. In general, **monographs** will be sent out to at least 2 external referees for evaluation.

   A final decision to publish can be made only on the basis of the complete manuscript, however a refereeing process leading to a preliminary decision can be based on a pre-final or incomplete manuscript.

   Volume Editors of **multi-author works** are expected to arrange for the refereeing, to the usual scientific standards, of the individual contributions. If the resulting reports can be

forwarded to the LNM Editorial Board, this is very helpful. If no reports are forwarded
or if other questions remain unclear in respect of homogeneity etc, the series editors may
wish to consult external referees for an overall evaluation of the volume.

5. Manuscripts should in general be submitted in English. Final manuscripts should contain
at least 100 pages of mathematical text and should always include

   - a table of contents;
   - an informative introduction, with adequate motivation and perhaps some historical
     remarks: it should be accessible to a reader not intimately familiar with the topic
     treated;
   - a subject index: as a rule this is genuinely helpful for the reader.
   - For evaluation purposes, manuscripts should be submitted as pdf files.

6. Careful preparation of the manuscripts will help keep production time short besides
ensuring satisfactory appearance of the finished book in print and online. After
acceptance of the manuscript authors will be asked to prepare the final LaTeX source
files (see LaTeX templates online: https://www.springer.com/gb/authors-editors/book-
authors-editors/manuscriptpreparation/5636) plus the corresponding pdf- or zipped ps-
file. The LaTeX source files are essential for producing the full-text online version of
the book, see http://link.springer.com/bookseries/304 for the existing online volumes
of LNM). The technical production of a Lecture Notes volume takes approximately 12
weeks. Additional instructions, if necessary, are available on request from lnm@springer.
com.

7. Authors receive a total of 30 free copies of their volume and free access to their book on
SpringerLink, but no royalties. They are entitled to a discount of 33.3 % on the price of
Springer books purchased for their personal use, if ordering directly from Springer.

8. Commitment to publish is made by a *Publishing Agreement*; contributing authors of
multiauthor books are requested to sign a *Consent to Publish form*. Springer-Verlag
registers the copyright for each volume. Authors are free to reuse material contained in
their LNM volumes in later publications: a brief written (or e-mail) request for formal
permission is sufficient.

Addresses:
Professor Jean-Michel Morel, CMLA, École Normale Supérieure de Cachan, France
E-mail: moreljeanmichel@gmail.com

Professor Bernard Teissier, Equipe Géométrie et Dynamique,
Institut de Mathématiques de Jussieu – Paris Rive Gauche, Paris, France
E-mail: bernard.teissier@imj-prg.fr

Springer: Ute McCrory, Mathematics, Heidelberg, Germany,
E-mail: lnm@springer.com